NEMATODE PESTS OF RICE

M. SHAMIM JAIRAJPURI
PhD, DSc, FZSI, FNASc, FLC, FIBiol
Director

and

QAISER H. BAQRI
PhD, DSc, FZSI
Deputy Director

Zoological Survey of India
'M' Block, New Alipore
Calcutta 700053, India

LONDON AND NEW YORK

First published 1992 by Westview Press

Published 2018 by Routledge
52 Vanderbilt Avenue, New York, NY 10017
2 Park Square, Milton Park, Abingdon, Oxon OX14 4RN

Routledge is an imprint of the Taylor & Francis Group, an informa business

ISBN 13: 978-0-367-00488-0 (hbk)

CIP data available upon request

ISBN 13: 978-0-367-15475-2 (pbk)

CONTENTS

Foreword: Dr. M.S. Swaminathan v
Preface vii

Introduction 1

Ecological Classification of Important Nematodes of Paddy 3
Systematic Position of Important Nemtodes of Paddy 3

Descriptions of Species

1. *Tylenchorhynchus annulatus* 5
2. *Tylenchorhynchus mashhoodi* 7
3. *Hoplolaimus indicus* 8
4. *Helicotylenchus dihystera* 10
5. *Helicotylenchus crenacauda* 10
6. *Helicotylenchus abunaamai* 13
7. *Pratylenchus coffeae* 13
8. *Pratylenchus thornei* 15
9. *Hirschmanniella oryzae* 16
10. *Hirschmanniella gracilis* 18
11. *Hirschmanniella mucronata* 20
12. *Hirschmanniella spinicaudata* 20
13. *Hirschmanniella caudacrena* 21
14. *Hirschmanniella belli* 21
15. *Hirschmanniella imamuri* 23

Rice Root Nematodes **23**

16. *Ditylenchus angustus* 25

Stem Nematode: Ufra **27**

17. *Meloidogyne javanica* 32
18. *Meloidogyne incognita* 32
19. *Meloidogyne graminicola* 35

Root Knot Nematodes **37**

20. *Heterodera oryzicola* 39
21. *Heterodera oryzae* 41

Cyst Nematodes **43**

22. *Criconemella onoensis* 43
23. *Criconemella rustica* 45
24. *Hemicriconemoides cocophillus* 46
25. *Aphelenchoides besseyi* 46

White-Tip Disease **48**

26. *Xiphinema insigne* 50
27. *Paratrichodorus porosus* 51

Key to Nematodes of Paddy 53
Other Plant Parasitic Nematodes from Rice Fields 56
Nematodes Found from Soil Around Roots of Rice 59

References 61

FOREWORD

The triple alliance of pests, pathogens and weeds takes a heavy toll on the yield of rice. Nematodes in particular can cause considerable damage and also present great difficulties in their eradication. Because of difficulties in identification, many nematode pests of rice had gone unnoticed in the past. It is in this context that the present book by Drs. M. Shamim Jairajpuri and Qaiser H. Baqri is a timely contribution. The book provides clear guidelines and information on the taxonomic identification on nematode pests, in addition to giving control and management schedules.

The only pathway open to us for producing more food for the use of present and future populations is productivity improvement through both a higher intensity of farming and higher yield per crop. It is in this context that the provision of an effective plant protection umbrella to the rice crop assumes urgency. I would therefore like to express the gratitude of all interested in rice research and production to Dr. Jairajpuri and Dr. Baqri for this labour of love.

M.S. Swaminathan Research Foundation
Madras, India

20 February 1991

Dr. M.S. Swaminathan
PhD, DSc, FNA, FRS

Former: Director, IARI, New Delhi
Director-General, ICAR, and
Secretary, Deptt. of Agric. Res. Ed., New Delhi
Director-General, IRRI, Philippines.

PREFACE

Rice is the most important cereal crop which is grown and consumed all over the world. The rice plant, *Oryza sativa* L., is cultivated in lowlands under flooded conditions, well drained upland fields or on the hills. The total area under rice cultivation in the world is about 140 million hectares. In India alone rice is grown on 37 million hectares and the total production is approximately 65 million tons, nearly one-third of our total grain production.

A large number of pests attack rice and some of these cause serious damage to the crop. The nematodes are considered very important pests of rice particularly after Butler (1913a) reported ufra disease from Bengal, India. In later years a number of other nematode species were found to be potential pests of rice and as a consequence significant work has been done on the symptomology, yield losses, control, etc. With the exception of white-tip and ufra diseases, the main difficulty in the identification of problems caused by nematodes is the absence of any clear-cut above-ground symptoms of their attack. Stunting, yellowing, curling of leaves, etc. which are generally associated with nematode infestation are also caused by other pests. Keeping in view the economic importance of nematode pests of rice, a sincere attempt has been made to provide information on important nematode species and nematode problems of rice. These are likely to be of use to students of nematology as also to the scientists involved in teaching, advisory and extension work. The aim of this book is to provide consolidated data on nematodes of rice, their identification, symptoms of diseases, yield losses, control measures, etc., with special reference to Indian subcontinent.

We are thankful to the CAB-International Institute of Parasitology, St. Albans, U.K., for granting us permission to use illustrations of nematodes published in CIH Descriptions of Plant-parasitic Nematodes and also to the Department of Nematology, Assam Agricultural University, Jorhat, for the use of photographs showing symptoms of disease due to nematode attack.

M. Shamim Jairajpuri
Qaiser H. Baqri

Zoological Survey of India
'M' Block, New Alipore
Calcutta 700 053, India.

Introduction

The nematodes like insects are a highly diversified group of animals and occur in all possible kinds of climatic conditions and habitats. For the existence and survival of nematodes, free-living as well as parasitic, either water or liquid medium is necessary. Those adapted to a terrestrial mode of life survive because of the moisture holding capability of the soil particles. In due course of evolution soil-inhabiting nematodes including those parasitic on plants have adjusted to live in conditions requiring a minimal quantity of water. The paddy crop is exceptional being a specialized habitat in the sense that the soil particles have very high water holding capacity. In deep-water rice the roots are virtually under water for most of the time. Though aquatic nematodes are adapted to thrive under such conditions because of their adaptability to low levels of oxygen and excessive water, for the plant-parasitic nematodes it is a return to ancestral conditions to which only a selected few have been able to adjust.

As compared to crops, such as, wheat, cotton and corn, our knowledge of nematodes parasitizing paddy is poor. This is due to the fact that paddy is grown mostly in the developing or underdeveloped areas of the world and it is in these countries that proper facilities and scientific personnel who could tackle the problem are lacking. According to estimates out of 346 million acres of rice grown in the world 263 million acres are infested with damaging levels of plant parasitic nematodes (Hollis & Keoboonrueng, 1984). It has roughly been estimated that the world-wide annual yield loss due to plant parasitic nematodes on rice ranges from 10-25%. In Asian countries, particularly the Indian subcontinent, more specifically the eastern regions of India including Bangladesh, rice is grown extensively and a large chunk of human population depends on it for its livelihood. The paddy crop though different from other crops like wheat because of its excessive water requirements, suffers damage from all kinds of pests, for example, insects, bacteria, fungi viruses and nematodes. Though our knowledge of nematode pests of rice and the damage caused by them is far from complete, yet the work done in India is fairly satisfactory and the same results could be taken as applicable in case of neighbouring countries like Bangladesh (Rao *et al.*, 1986; Rao, 1988).

The studies so far carried out in India show that several groups of nematodes are present in the paddy fields including those which are specific parasites of the roots, stem and leaves of rice plants. The various groups that are found may be classified as free-living, predaceous and plant-parasites. Though only the last group merits consideration from the economic point of view, yet the predaceous nematodes in particular and the free-living ones to some extent should not be ignored as they also have an important role in the overall ecosystem as well as in pest management programmes. The predaceous nematodes, particularly those belonging to the Order Mononchida (popularly called mononchs) feed voraciously on both, free-living and plant-parasitic nematodes and hence are important tools in the biological control of phytophagous nematodes.

The plant-parasitic nematodes that are associated with paddy belong to several groups with different kinds, characteristics, feeding habits, behaviour, pathogenecity, etc. Both the major groups, that is, those belonging to the Order Tylenchida (popularly called tylenchs) and the Order Aphelenchida (popularly known as aphelenchs) are found, though the former outnumber the latter by a vast majority. Some of these nematodes feed as temporary ectoparasites of roots (*Tylenchorhynchus, Helicotylenchus*), while others may remain often permanently attached to the roots (*Criconemella, Caloosia*). A number of species of nematodes belonging to *Hirschmanniella, Pratylenchus* and *Hoplolaimus* enter the root system and feed as migratory endoparasites but these may also be recognised as migratory ectoparasites. The cyst forming nematodes of the genus *Heterodera* and the root knot nematodes belonging to *Meloidogyne* are examples of sedentary endoparasites of root, though the former is regarded as semi-endoparasite by some because of the fact that the major part of the body of adult female lies outside the root, only its neck and the head region remaining inside for feeding purposes. Apart from feeding on the roots (underground parts) of the rice plants, two important species of nematodes, namely, *Ditylenchus angustus* and *Aphelenchoides besseyi* feed on the above-ground parts. *D. angustus*, the causative agent of '*urfa*' or '*Dak pora*' disease, particularly of submerged rice culture and deep-water paddy, attack the stem of the plants. *Aphelenchoides besseyi* is known to cause the widespread and well-known 'white tip' disease of rice. The nematodes feed on leaves causing reduction in chlorophyll content which turn pale yellow to white especially in the upper 2-5 cm (Ichinohe, 1972).

ECOLOGICAL CLASSIFICATION OF IMPORTANT NEMATODES OF PADDY

1. MIGRATORY ENDOPARASITIC NEMATODES OF ROOTS

   (i) Rice Root Nematodes
   Hirschmanniella spp.

   (ii) Lance Nematodes
   Hoplolaimus spp.

   (iii) Root Lesion Nematodes
   Pratylenchus spp.

2. ECTOPARASITIC NEMATODES OF ROOTS

   (i) *Tylenchorhynchus* spp.

   (ii) *Helicotylenchus* spp.

   (iii) *Criconemella* spp.

3. ECTOPARASITIC NEMATODES OF FOLIAGE

   (i) White-tip Nematodes
   Aphelenchoides besseyi

   (ii) Stem Nematode
   Ditylenchus angustus

4. SEDENTARY ENDOPARASITES OF ROOTS

   (i) Root knot Nematodes
   Meloidogyne spp.

   (ii) Cyst Nematodes
   Heterodera spp.

SYSTEMATIC POSITION OF IMPORTANT NEMATODES OF PADDY

ORDER TYLENCHIDA Thorne, 1949
SUBORDER TYLENCHINA Chitwood in Chitwood & Chitwood, 1950
FAMILY TYLENCHORHYNCHIDAE Eliava, 1964

Tylenchorhynchus Cobb, 1913
T. annulatus (Cassidy, 1930) Golden, 1971
T. mashhoodi Siddiqi & Basir, 1959

FAMILY HOPLOLAIMIDAE Filipjev, 1934

Hoplolaimus Daday, 1905
H. indicus Sher, 1963
Helicotylenchus Steiner, 1945
H. dihystera (Cobb, 1893) Sher, 1961
H. crenacauda Sher, 1966
H. abunaamai Siddiqi, 1971

FAMILY PRATYLENCHIDAE Thorne, 1949

Pratylenchus Filipjev, 1936
P. coffeae (Zimmermann, 1893) Filipjev & Sch. Stek., 1941
P. thornei Sher & Allen, 1953
Hirschmanniella Luc & Goodey, 1963
H. oryzae (van Breda de Haan, 1902) Luc & Goodey, 1963
H. gracilis (De Man, 1880) Luc & Goodey, 1963
H. mucronata (Das, 1960) Luc & Goodey, 1963
H. spinicaudata (Sch. Stek., 1944) Luc & Goodey, 1963
H. caudacrena Sher, 1968
H. belli Sher, 1968
H. imamuri Sher, 1968

FAMILY ANGUINIDAE Nicoll, 1935

Ditylenchus Filipjev, 1936
D. angustus (Butler, 1913) Filipjev, 1934

FAMILY MELOIDOGYNIDAE Skarbilovich, 1959

Meloidogyne Goeldi, 1892
M. javanica (Trueb, 1885) Chitwood, 1949
M. incognita (Kofoid & White, 1919) Chitwood, 1949
M. graminicola Golden & Birchfield, 1965

FAMILY HETERODERIDAE Filipjev & Sch. Stek., 1941

Heterodera Schmidt, 1871
H. oryzicola Rao & Jayaprakash, 1978
H. oryzae Luc & Berdon, 1961

SUBORDER CRICONEMATINA Siddiqi, 1980

FAMILY CRICONEMATIDAE Taylor, 1936

Criconemella De Grisse & Loof, 1965
C. onoensis (Luc, 1959) Luc & Raski, 1981
C. rustica (Micoletzky, 1915) Luc & Raski, 1981
Hemicriconemoides Chitwood & Birchfield, 1957
H. cocophillus (Loos, 1949) Chitwood & Birchfield, 1957

ORDER APHELENCHIDA Siddiqi, 1980

FAMILY APHELENCHOIDIDAE Skarbilovich, 1947

Aphelenchoides Fischer, 1894
A. besseyi Christie, 1942

ORDER DORYLAIMIDA Pearse, 1942

FAMILY XIPHINEMATIDAE Dalmasso, 1969

Xiphinema Cobb, 1913
X. insigne Loos, 1949

ORDER TRIPLONCHIDA Cobb, 1920

FAMILY TRICHODORIDAE Thorne, 1935

Paratrichodorus Siddiqi, 1974
P. porosus (Allen, 1957) Siddiqi, 1974

DESCRIPTIONS OF SPECIES

1. TYLENCHORHYNCHUS ANNULATUS (CASSIDY, 1930) GOLDEN, 1971

(Fig. 1, A-F)

Syn. *Tylopharynx annulatus* Cassidy, 1930
Anguillulina annulata (Cassidy, 1930), Goodey, 1932
Chitinotylenchus annulatus (Cassidy, 1930), Filipjev, 1936
Tylenchorhynchus martini Fielding, 1956

Measurements

Females: L = 0.66-0.72 mm; a = 31-35; b = 5.0-5.6; c = 15-16; c′ = 3.1-3.6; V = 54-56.

Description

Female: Body slightly ventrally curved upon fixation. Cuticle transversely striated, 1.5-2.0 μm apart. Longitudinal striae absent. Lateral fields marked with four incisures, outer incisures crenate. Lip region rounded, bearing three transverse striae. Cephalic framework lightly sclerotized. Stylet 17-18 μm long with rounded basal knobs having flattened to convex anterior surfaces. Oesophagus typical of the genus. Excretory pore opposite base of isthmus. Female reproductive system amphidelphic, spermatheca non-functional. Tail elongate, subcylindrical, 3.1-3.6 anal body-widths long, almost straight to slightly arcuate ventrally, marked by 18-22 striae ventrally with broadly rounded unstriated terminus. Phasmids located in the anterior half of tail.

Male: Absent.

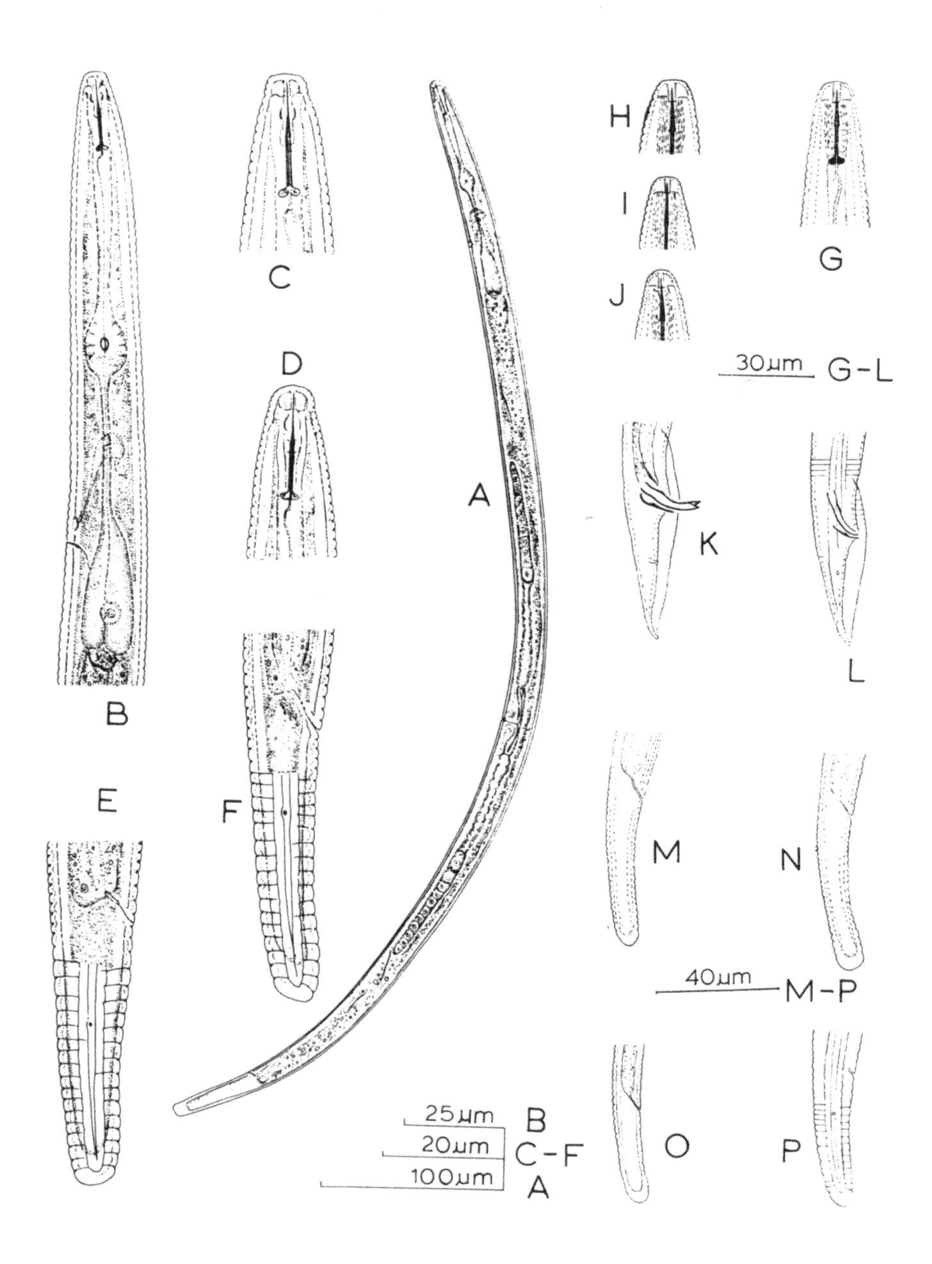

Fig. 1: A-F: *Tylenchorhynchus annulatus* (Cassidy, 1930) Golden, 1971 (after Siddiqi, 1976). A – Entire female; B – Oesophageal region; C & D – Head ends; E & F – Female tails. G-P: *Tylenchorhynchus mashhoodi* Siddiqi & Basir, 1959 (after Baqri & Jairajpuri, 1970). G-J – Head ends; K-L – Male tails; M-P – Female tails.

Remarks: *Tylenchorhynchus annulatus* is widely associated in tropical and subtropical regions with rice, sugarcane and grasses. Timm & Ameen (1960) have also reported this species from rice fields in Bangladesh. Joshi & Hollis (1976) have noted that swarming populations in greenhouse pots can cause significant reduction in root and shoot length and weight of rice. However, Hollis & Keoboonrueng (1984) have reported that extremely high rice yields have also been obtained in the presence of high population of *T. annulatus* in Lousiana and Texas. According to them the swarming incidence in the field is very low and as such the species may not cause damage significantly. Apart from chemical control, weed control is necessary in order to check *T. annulatus* populations. Siddiqi (1976) has provided a detailed redescription of this species.

2. **TYLENCHORHYNCHUS MASHHOODI** SIDDIQI & BASIR, 1959 (Fig. 1, G-P)

Syn. *Tylenchorhynchus dactylurus* Das, 1960
T. digittatus Das, 1960
T. crassicaudatus Williams, 1960
T. elegans Siddiqi, 1961
T. zeae Sethi & Swarup, 1968

Measurements

Females: L = 0.54-0.72 mm; a = 28-37; b = 4.4-5.8; c = 13-18; V = 52-58.

Males: L = 0.57-0.66 mm; a = 27-35; b = 4.5-5.6; c = 14- 17; T = 36-50.

Description

Female: Body ventrally curved in posterior half of its length upon fixation. Cuticle marked with distinct transverse striae, 1-2 μm apart. Longitudinal striations absent. Lateral fields marked by four incisures. Lip region continuous with body contour, bearing three to four annules. Cephalic framework lightly sclerotized. Stylet 16-19 μm long, with rounded basal knobs which may be slightly flattened anteriorly. Oesophagus typical of the genus. Median oesophageal bulb 44-50% of oesophageal length from anterior end. Female reproductive system amphidelphic, spermatheca functional. Tail cylindrical to subcylindrical in shape, slightly narrowing behind anus, rounded to bluntly rounded unstriated terminus, marked with 14-29 striae ventrally, and 2.5-4.0 anal body-widths long. Phasmids in anterior half (28-37%) of tail.

Males: Similar to female in general morphology. Spicules 18-22 μm long medially. Gubernaculum 10-15 μm long. Tail elongate-conoid with acute or subacute terminus, 2.3-3.0 anal body-widths long. Phasmids in the anterior half (31-50%) of tail.

Remarks: The species has a wide range of hosts and distribution. In India it is widely distributed in paddy fields and is regarded as a potential pest. Baqri &

Jairajpuri (1970) have redescribed the species along with its intraspecific variations.

3. HOPLOLAIMUS INDICUS SHER, 1963

(Fig. 2)

Syn. *Basirolaimus indicus* (Sher, 1963) Shamsi, 1979
Hoplolaimus arachidis Maharaju & Das, 1982

Measurements

Females: L = 1.02-1.40 mm; a = 22-36; b = 8.4-9.1; b′ = 7.0-8.1;
c = 45-74; c′ = 0.66-0.68; V = 50-59; o = 10-18.

Males: L = 0.94-1.30 mm; a = 26-36; b = 8.9-12.0; b′ = 6.2=9.0;
c = 31-38; o = 10-16.

Description

Female: Body ventrally curved upon fixation. Cuticle coarsely annulated. Lateral fields marked by single incisure or two to three incomplete broken incisures. Anterior phasmid about 24-44% from anterior end on right side and posterior phasmid at 76-86% from anterior end on left. Lip region hemispheroid marked by three to four annules, basal annule with six to 12 longitudinal striations. Cephalic framework strongly sclerotized. Stylet robust, 30-34 μm long; basal knobs usually provided with one to three forwardly directed processes. Median oesophageal bulb spheroid, with well-developed valvular apparatus. Oesophageal glands overlapping intestine dorsally and laterally, with six nuclei. Excretory pore located anterior to the level of oesophago-intestinal junction. Female reproductive system amphidelphic. Epiptygma single or double. Spermatheca filled with sperm. Intestine overlapping rectum. Tail round, with eight to 13 annules.

Male: Similar to female in general morphology. Spicules arcuate and cephalated, 37-48 μm long. Gubernaculum 12-20 μm long. Tail conoid. Bursa terminal.

Remarks: *Hoplolaimus indicus* is a widely distributed species in India. Bhubaneswar Centre of All India Coordinated Research Project (sponsored by ICAR) on nematode pests has reported that 80% samples from rice fields in Orissa state were found infested by this species. ZSI nematologists have found this species to be a potential pest of paddy crop in West Bengal. Khan & Chawla (1975) have provided detailed description of *H. indicus*.

Benerjee & Banerjee (1966) have reported severe damage to paddy due to *H. indicus* in West Bengal. Ramana & Rao (1978) have estimated the loss in yield due to these nematodes up to 20%. All stages of the nematode commonly occur inside the root tissues but the fourth stage juveniles and adults are more abundant in soil indicating thereby that they move out of the plant tissues into the soil towards maturity. The leaves of the infested plants become yellow and curled. Fallowing for three months brings down the population of nematodes

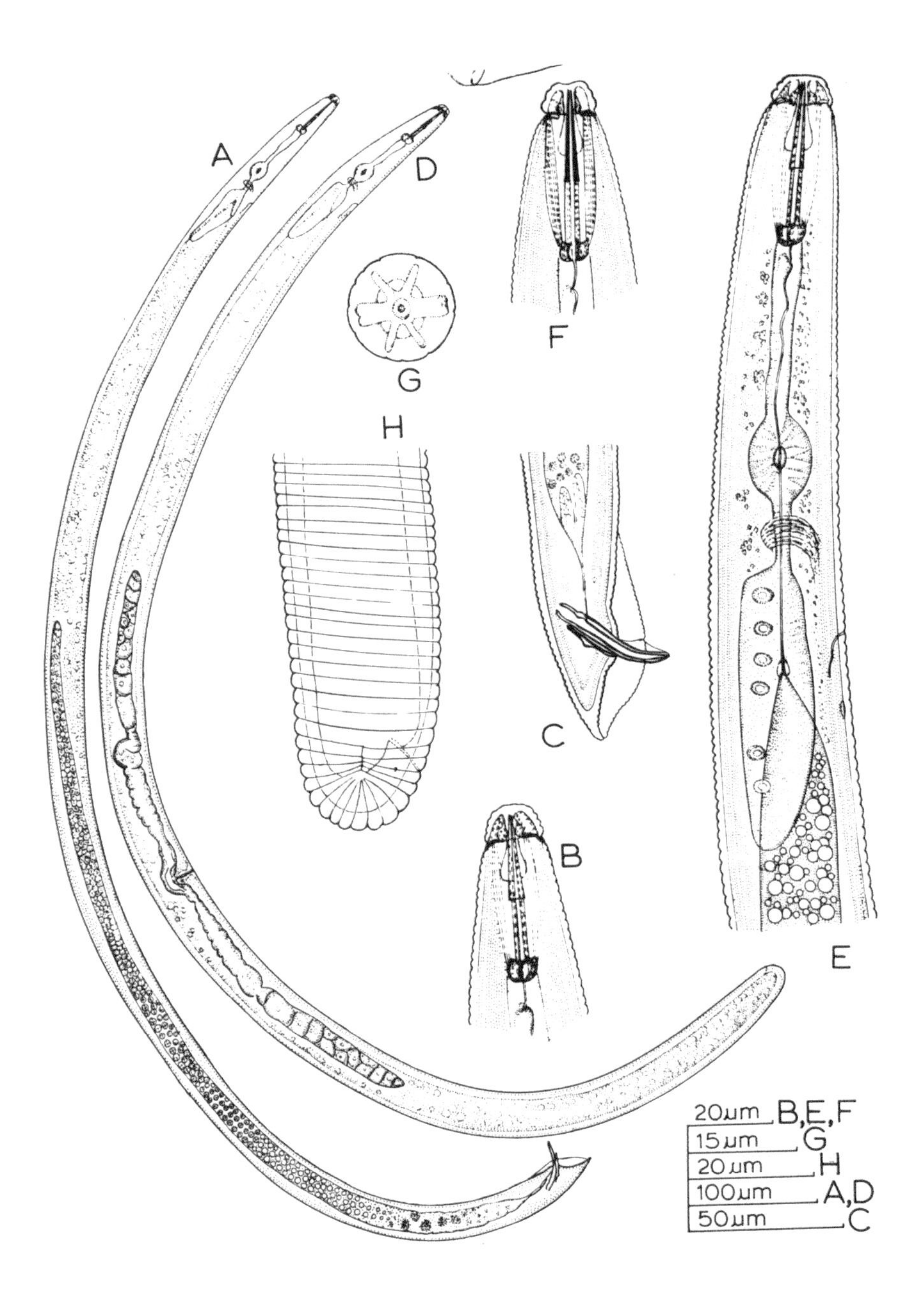

Fig. 2: *Hoplolaimus indicus* Sher, 1963 (after Khan & Chawla, 1975). A – Entire male; B – Male head end; C – Male tail; D – Entire female; E – Female oesophageal region; F – Female head end; G – *en-face* view; H – Female tail (superficial view).

significantly (Chawla, 1972). The preplanting root dip treatment of seedlings by diazinon or DBCP @ 500 ppm for 10 minutes has not only been found significantly effective in reducing *H. indicus* populations but also in suppressing its development inside the rice roots. The application of diazinon and dimethoate @ 1 kg a.i./ha in the field is effective against these nematodes.

4. HELICOTYLENCHUS DIHYSTERA (COBB, 1893) SHER, 1961

(Fig. 3, A-D)

Syn. *Tylenchus dihystera* Cobb, 1893
T. olae Cobb, 1906
T. spiralis Cassidy, 1930
Helicotylenchus nanus Steiner, 1945
H. crenatus Das, 1960
H. punicae Swarup & Sethi, 1968

Measurements

Females: L = 0.50-0.86 mm; a = 21-34; b = 4.5-6.4; b′ = 3.7-5.2; c = 34-65; c′ = 1.0-1.3; V = 60-66; o = 36-49.

Description

Female: Body spiral. Cuticle with distinct transverse striae. Lip region continuous with body, hemispherical bearing four to five annules. Cephalic framework conspicuous. Lateral fields marked by four incisures, not areolated. Stylet 24-27 μm long, basal knobs with concave or indented anterior surfaces. Oesophagus with overlapping glands, longest overlap is ventral. Female reproductive system amphidelphic. Spermatheca rounded, offset, without sperm. Tail dorsally convex-conoid, usually with slight ventral projection, sometimes only a narrow terminus (Baqri & Ahmad, 1984). The fusion of inner incisures at tail tip U-shaped, V-shaped or rarely Y-shaped.

Male: Extremely rare.

Remarks: This is a cosmopolitan and widely distributed species associated with many host plants including rice. Baqri & Ahmad (1984) have discussed the allometric and morphometric variations in *H. dihystera.*

5. HELICOTYLENCHUS CRENACAUDA SHER, 1966

(Fig. 4)

Syn. *Helicotylenchus pteracercus* Singh, 1971
H. indentatus Chaturvedi & Khera, 1979
? *H. parateracercus* Sultan, 1981

Measurements

Females: L = 0.50-0.70 mm; a = 20-29; b = 4.7-6.3; b′ = 4.0-5.2; c = 30-44; c′ = 0.9-1.4; V = 60-66; o = 33-40.

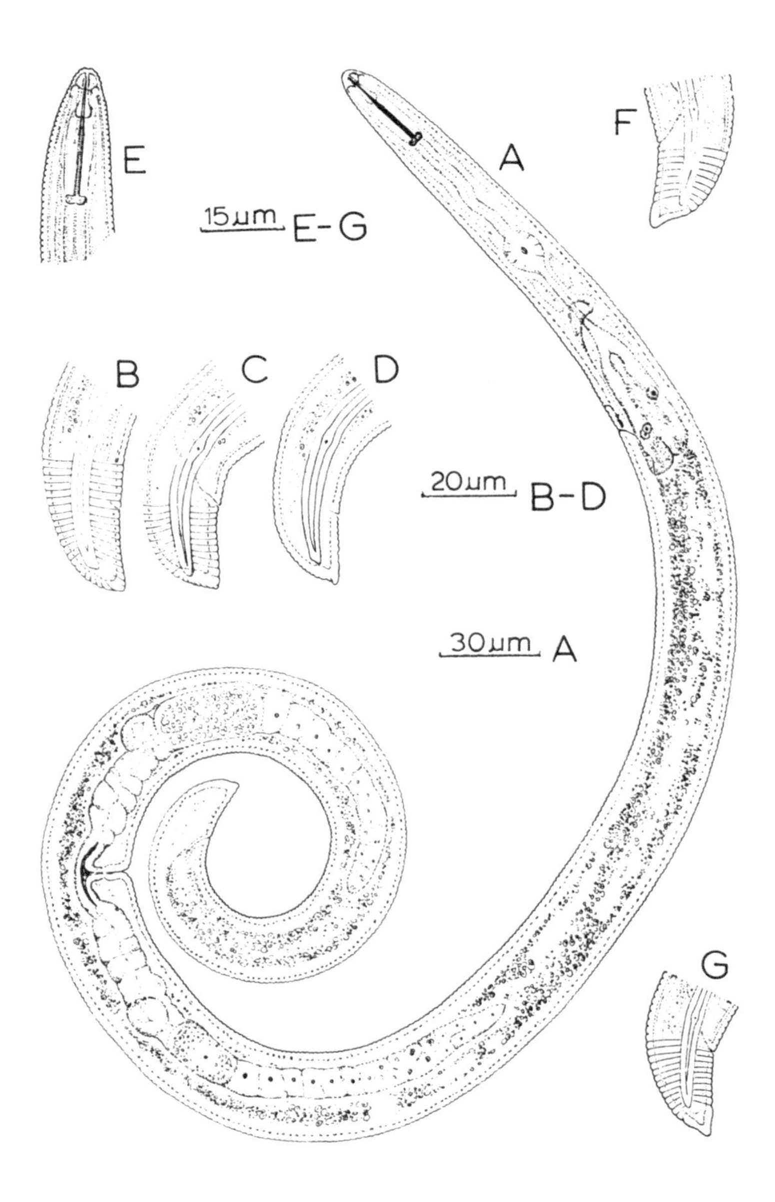

Fig. 3: A-D: *Helicotylenchus dihystera* (Cobb, 1893) Sher, 1961 (after Siddiqi, 1972). A – Entire female; B-D – Female tails. E-G: *Helicotylenchus abunaamai* Siddiqi, 1972 (after Siddiqi, 1972). E – Head end; F & G – Female tails.

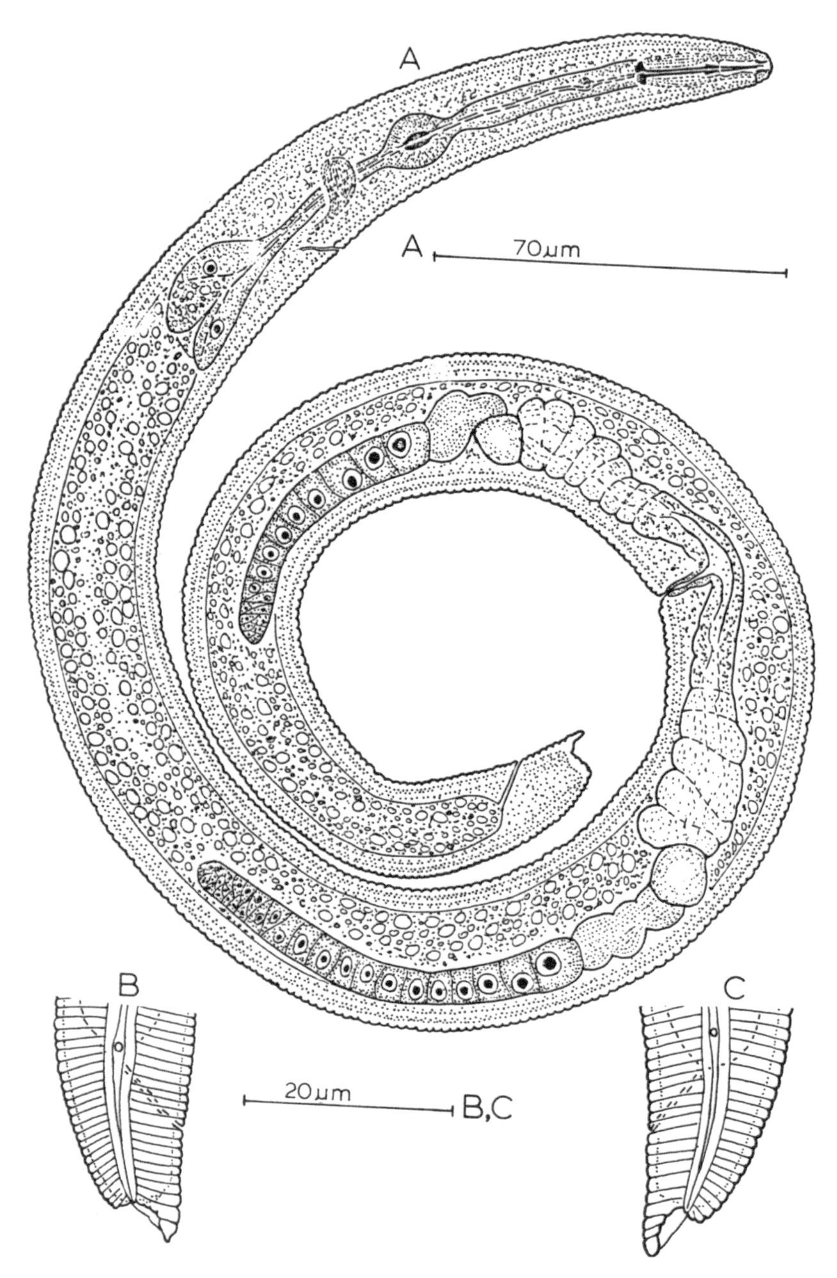

Fig. 4: *Helicotylenchus crenacauda* Sher, 1966 (after Baqri & Ahmad, 1984). A – Entire female; B & C – Female tails.

Description

Female: Body spirally curved upon fixation. Cuticle transversely striated. Lateral fields marked by four incisures, the inner incisures fused near the middle of tail. Lip region continuous, broadly rounded (80%) to truncated (20%), marked by four to five annules. Stylet 24-27 μm long; basal knobs flattened (30%) or indented (70%) anteriorly. Oesophagus typical of the genus. Female reproductive system amphidelphic. Spermatheca without sperm. Tail with a well-developed ventral projection and a characteristic indentation at dorsal terminal part, the ventral projection enveloped in cuticular fold originating from the terminal region of the lateral fields. Phasmids preanal.

Male: Not found.

Remarks: *H. crenacauda* has been found widely distributed in many districts of West Bengal (India) and Bangladesh. Baqri & Ahmad (1984) have provided the allometric and morphometric variations in this species.

6. **HELICOTYLENCHUS ABUNAAMAI** SIDDIQI, 1972

(Fig. 3, E-G)

Measurements

Females: L = 0.50-0.63 mm; a = 25-29; b = 5.5-6.7; b′ = 4.5-5.1; c = 33-44; c′ = 1.1-1.4; V = 59-65; o = 41-45.

Description

Female: Body spirally curved. Lateral fields with four smooth incisures. Lip region hemispheroid, broadly rounded, continuous with body, marked by four annules. Stylet 21-22 μm long, with flattened to slightly concave anterior surface. Oesophagus typical of the genus. Female reproductive system amphidelphic. Spermatheca without sperm. Tail ventrally convex and dorsally concave; with a smooth narrow hemispheroid terminus giving the appearance of a ventral projection, rarely rounded. Phasmids preanal.

Male: Not found.

Remarks: This species has been found in 80% paddy fields in Orissa state. Padhi & Das (1986) have studied the life cycle of *H. abunaamai* and noted that feeding on host is essential for the development of all stages.

Helicotylenchus dihystera, H. caudacrena and *H. abunaamai* attack many plants and cause damage but on rice, though they occur in very large populations, the extent of damage is not known.

7. **PRATYLENCHUS COFFEAE** (ZIMMERMANN, 1898) FILIPJEV & SCH. STEK., 1941

(Fig. 5, A-H)

Syn. *Pratylenchus musicola* (Cobb, 1919)
P. mahogani (Cobb, 1920)
P. pratensis apud Yokoo, 1956

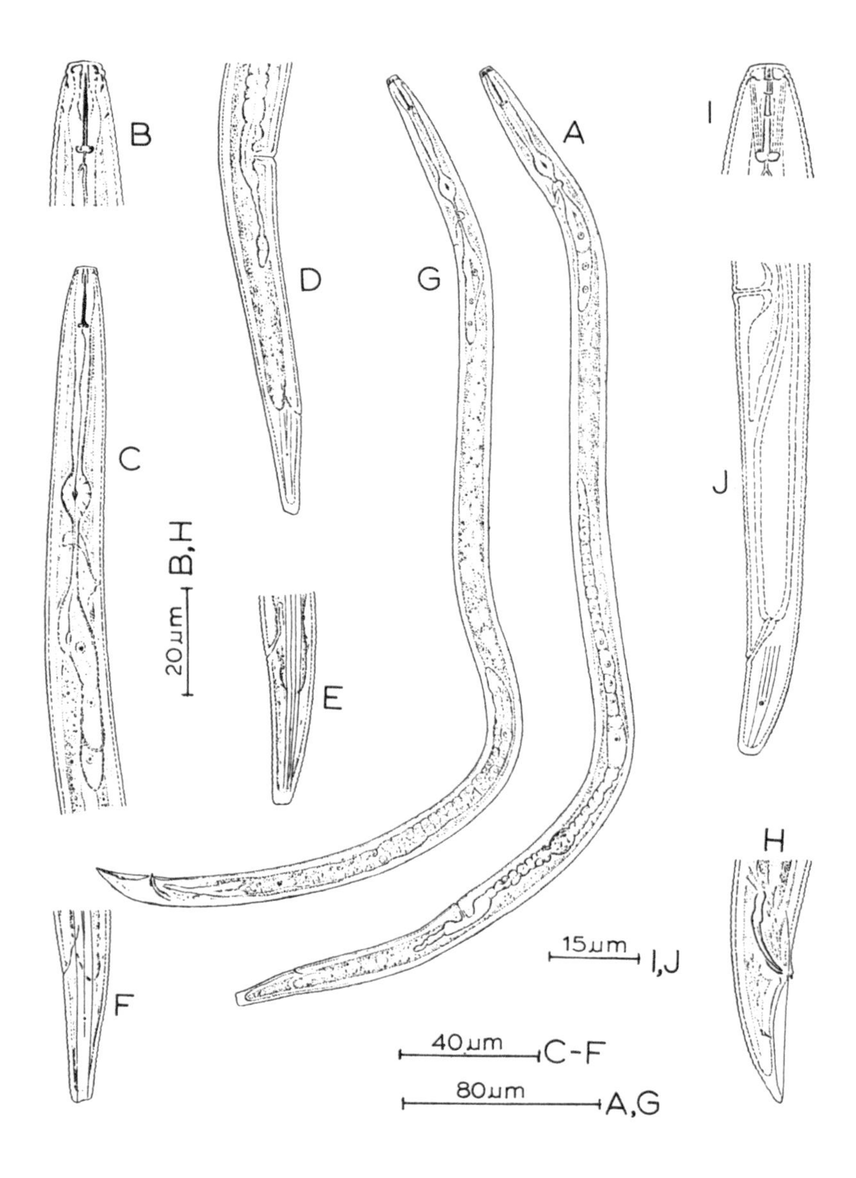

Fig. 5: A-H: *Pratylenchus coffeae* (Zimmermann, 1898) Filipjev & Sch.Stek., 1941 (after Siddiqi, 1972). A – Entire female; B – Head end; C – Oesophageal region; D – Posterior region of female; E & F – Female tails; G – Entire male; H – Male tail. I-J: *Pratylenchus thornei* Sher & Allen, 1953 (after Thorne 1961). I – Head end; J – Posterior region of female.

Measurements

Females: L = 0.37-0.68 mm; a = 17.7-30.5; b = 5.0-7.8; c = 13.7-23.9; V = 75.8-84.2.

Males: L = 0.40-0.56 mm; a = 23.8-31.8; b = 5.9-7.7; c = 17.6-23.3; T = 37-58.

Description

Female: Body almost straight. Cuticle distinctly transversely striated. Lateral fields marked by four incisures. Lip region low, marked by two annules. Stylet stout, 15-18 μm long, with well-developed basal knobs. Oesophagus typical of the genus. Female reproductive system prodelphic. Spermatheca filled with sperm, oval. Posterior uterine branch variable in length and structure, sometimes up to 50 μm long. Tail tapering slightly; tip broadly rounded, truncated or indented.

Male: Similar to female in general morphology except for reproductive system and tail shape. Spicules slender, 16-20 μm long. Gubernaculum 4-7 μm long.

Remarks: This is a widely distributed species in pantropical and subtropical countries including India and Bangladesh.

8. **PRATYLENCHUS THORNEI** SHER & ALLEN, 1953
(Fig. 5, I-J)

Measurements

Females: L = 0.40-0.70 mm; a = 25-36; b = 5.4-8.3; c = 18.6-25.1; V = 73-80.

Males: L = 0.49 mm; a = 29; b = 6.2; c = 20.3; T = ?

Description

Female: Body open 'C' shaped unlike other species of *Pratylenchus*, tapering posterior to vulva. Cuticle finely striated. Lateral fields marked by four incisures, the outer ones are slightly crenate. Lip region continuous, comparatively higher, marked by three annules; labial sclerotization extending backwards on two to three body annules. Stylet 15-19 μm long. Female reproductive system prodelphic. Spermatheca without sperm. Tail tip broadly rounded to truncate, smooth.

Male: Very rare, similar to female in general morphology, except for its reproductive system. Spicules slender, 21 μm long.

Remarks: Low soil moisture content (30-50%) favours high population build up of lesion nematodes, *Pratylenchus* spp. Consequently, those areas with comparatively low rainfall are infested with these nematodes, for example, Bihar, Uttar Pradesh, Maharashtra and Orissa. High populations have been recorded also from upland areas such as Quilon, Trivandrum and Cuttack.

The reduction in chlorophyll, number of ear-heads, etc., may cause significant grain yield losses due to *Pratylenchus* species, depending upon the intensity of invasion and environmental factors. Rao et al. (1986) have reported the loss of yield up to 53% due to lesion nematodes.

The non-host crops namely, blackgram, oats, wheat and greengram may be grown in rotation with rice to bring down the populations of *Pratylenchus* spp. The use of carbofuran granules in the field @ 1 kg a.i./ha when the crop is 30 days old reduces the population levels significantly.

9. **HIRSCHMANNIELLA ORYZAE** (VAN BREDA DE HAAN, 1902) LUC & GOODEY, 1963

(Fig. 6, A-J)

Syn. *Tylenchus oryzae* van Breda de Haan, 1902
T. apapillatus Imamura, 1931
Anguillulina oryzae (van Breda de Haan) T. Goodey, 1932
Rotylenchus oryzae (van Breda de Haan) Filipjev & Sch.Stek., 1941
Radopholus oryzae (van Breda de Haan) Thorne, 1949
Hirschmannia oryzae (van Breda de Haan) Luc & Goodey, 1962
Hirschmanniella nana Siddiqi, 1966
nec Tylenchus oryzae Soltwedel, 1889 (= *nomen nudum*)

Measurements

Females: L = 1.14-1.63 mm; a = 50-67; b = 8.8-12.1; b′ = 4.5-7.2; c = 15-19; c′ = 4.3-5.5; V = 50-55; o = 15-19.

Males: L = 1.01-1.40 mm; a = 52-61; b = 9.1-11.3; b′ = 4.6-5.7; c = 16-18; o = 13-18.

Description

Female: Body straight or slightly curved ventrally. Cuticle transversely striated; 1.2-2.0 μm apart on mid-body. Lateral fields marked with four incisures of which outer ones are crenate, sometimes incomplete areolations may occur on tail. Lip region continuous with body, low, flattened with rounded edges, marked by three to four annules. Stylet robust, 16-19 μm, with rounded basal knobs. Oesophagus typical of the genus. Female reproductive system amphidelphic. Spermatheca with sperm. Intestine not overlapping rectum. Tail elongate conoid, 4.3 to 6.5 anal body-widths long, with mucronate terminus. Phasmids located in the posterior half of tail.

Male: Similar to female in general morphology except in reproductive system and tail shape. Spicules cephalated, slightly arcuate, 18-28 μm long. Gubernaculum 7-9 μm long. Bursa subterminal. Tail with a pointed ventral mucro.

Remarks: *H. oryzae* is widely distributed and considered a key pest of rice in far East and tropical countries. On rice, it is usually sympatric with one or more species of *Hirshmanniella*. Siddiqi (1973) has provided a detailed redescription of this species.

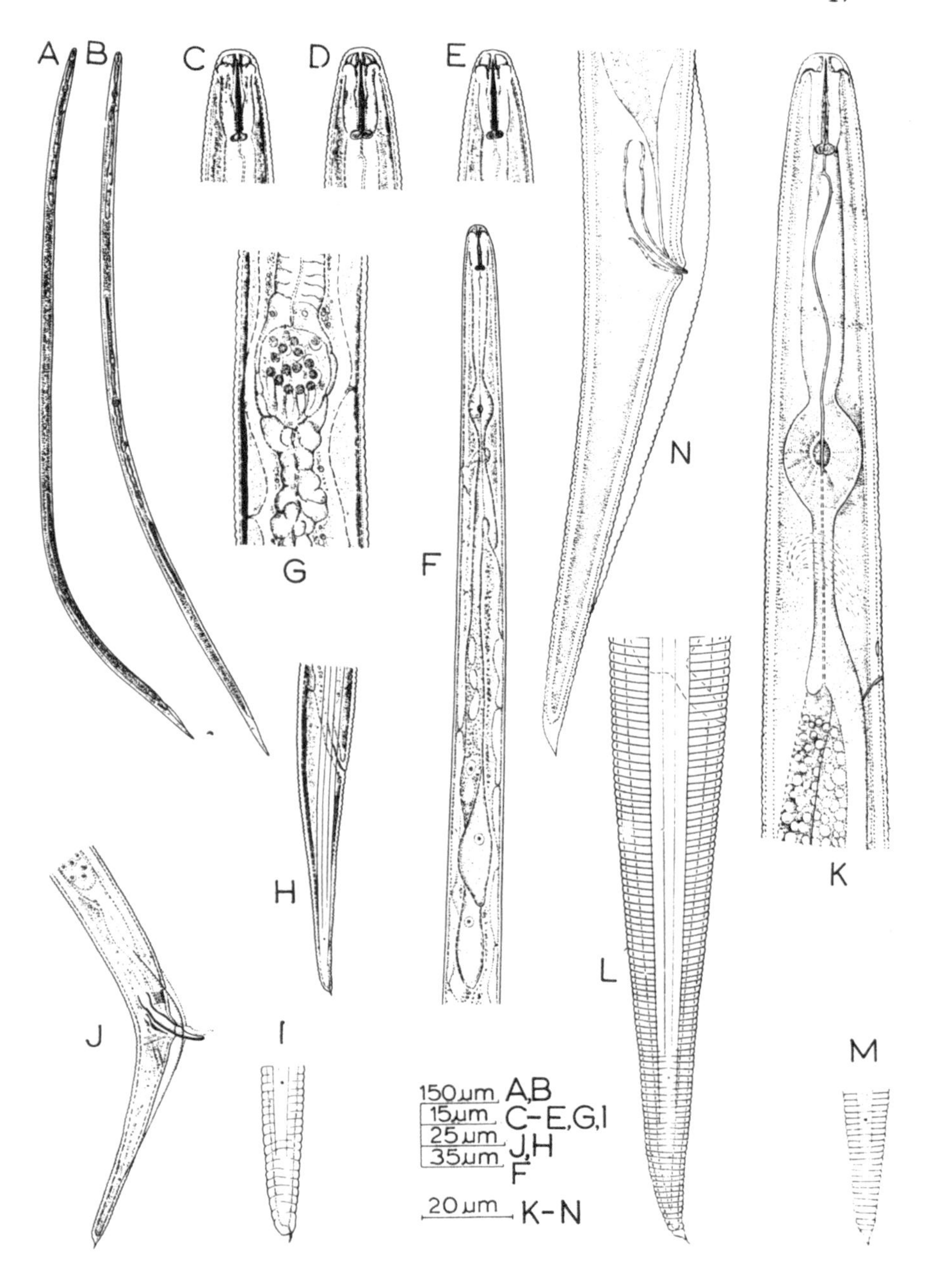

Fig. 6: A-J: *Hirschmanniella oryzae* (van Breda de Haan, 1902) Luc & Goodey, 1963 (after Siddiqi, 1973). A – Entire male; B – Entire female; C-E – Head ends; F – Oesophageal region; G – Spermatheca with sperm; H – Female tails; I – Female tail end; J – Male tail. K-N: *Hirschmanniella gracilis* (de Man, 1880) Luc & Goodey, 1963 (after Sher, 1968). K – Oesophageal region; L – Female tail; M – Female tail end; N – Male tail.

10. **HIRSCHMANNIELLA GRACILIS** (DE MAN, 1880) LUC & GOODEY, 1963

(Fig. 6, K-N)

Syn. *Tylenchus gracilis* de Man, 1880
Anguillulina gracilis (de Man) T. Goodey, 1932
Tylenchorhynchus gracilis (de Man) Micoletzky, 1925
Rodopholus gigas Andrássy, 1954
Hirschmannia gracilis (de Man) Luc & Goodey, 1962

Measurements

Females: L = 1.43-1.96 mm; a = 42-62; b = 11.6-17; b′ = 4.5-8.7; c = 16-21; c′ = 3.4-5.9; V = 48-54; o = 13-20.

Males: L = 1.22-1.182 mm; a = 42-64; b = 10.4-13.7; b′ = 4.1-7.2; c = 15.5-22.5; c′ = 3.4-6.0; o = 13-21.

Description

Female: Body slightly curved ventrally. Cuticle marked by transverse striae, 1.2-2.0 μm apart. Lateral fields marked by four incisures, outer ones crenate, usually not areolated but incomplete areolations may occur in tail region. Lip region continuous with body, apex usually flattened, marked by three to five annules. Stylet robust, 21-24 μm long with rounded basal knobs. Oesophagus typical of the genus. Female reproductive system amphidelphic. Spermatheca filled with sperm. Tail elongate conoid, 3.5 to 6.0 anal body-widths long, terminus with a pointed ventral projection. Phasmids located in posterior half of tail, 21-41% of tail length from terminus.

Male: Similar to female in general shape and morphology except reproductive system and tail shape. Spicules 28-38 μm long when measured along median line. Gubernaculum 9-14 μm long. Bursa subterminal. Tail slightly ventrally curved with terminus bearing a pointed ventral projection.

Remarks: Sivakumar & Khan (1982) have separated *H. gracilis* with its closely related species *H. oryzae* on the following characters: Lateral fields completely or incompletely areolated and excretory pore opposite oesophago-intestinal valve in *H. gracilis* against areolations absent or incomplete, and excretory pore posterior to oesophago-intestinal valve in *H. oryzae*. However, Dey & Baqri (1985) while giving the allometric and morphometric variations in *H. gracilis* have concluded that the position of excretory pore is a moderately variable character which is significantly correlated to the body length and the visibility of areolations depends upon fixation techniques. They (l.c.) have further concluded that the stylet length should be used as a key character for separating these two species (stylet length 20-24 μm in *H. gracilis* as against 15-19 μm in *H. oryzae*).

H. gracilis is a widely distributed species in the U.S.A., Canada, Netherlands, Germany, India and many other countries. Baqri et al. (1983) have reported this species as a key pest of rice in West Bengal.

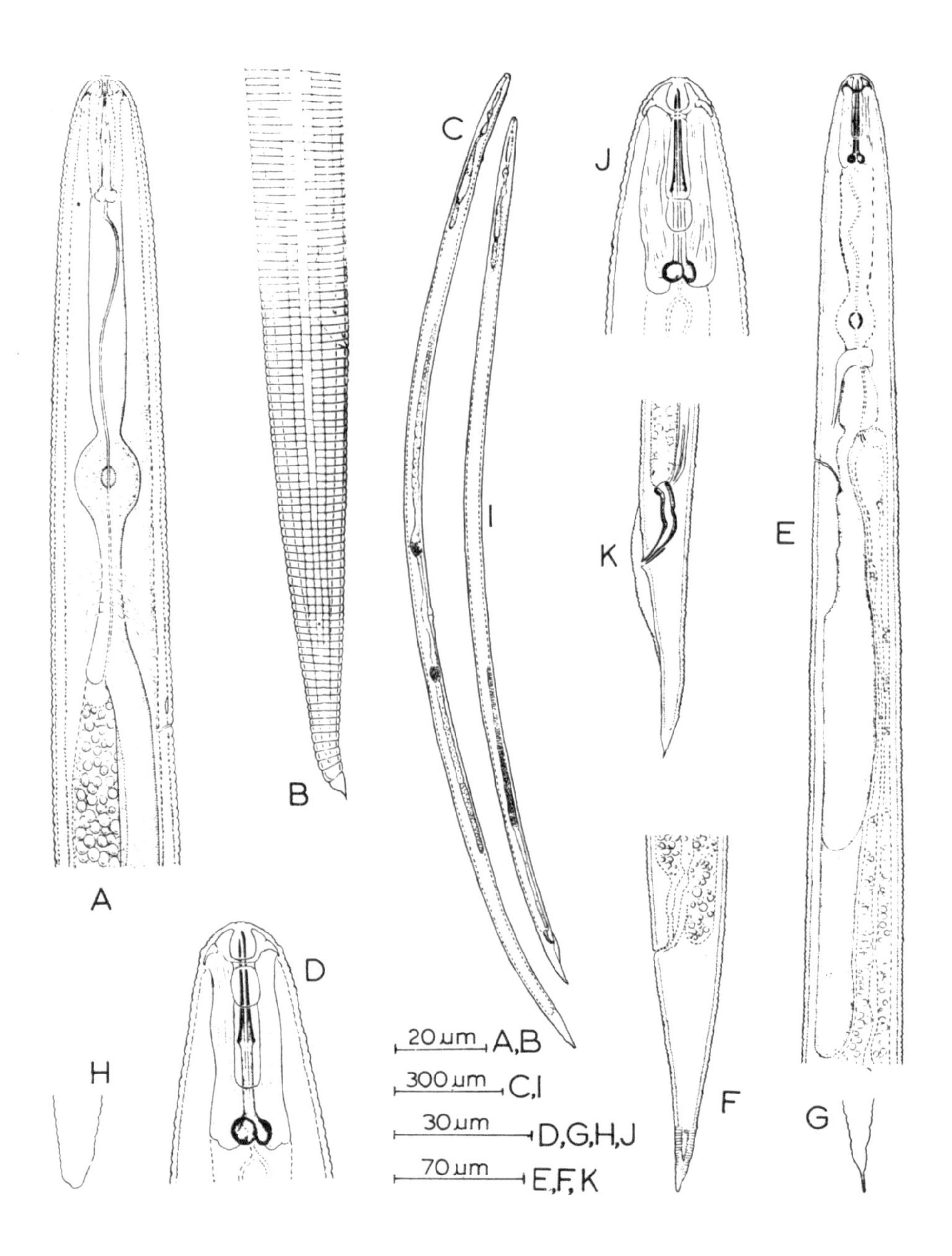

Fig. 7: A-B: *Hirschmanniella mucronata* (Das, 1960) Luc & Goodey, 1963 (after Sher, 1968). A – Oesophageal region; B – Female tail. C-K: *Hirschmanniella spinicaudata* (Sch. Stek., 1944) Luc & Goodey, 1963 (after Luc & Fortuner, 1975). C – Entire female; D – Female head end; E – Oesophageal region; F – Female tail; G & H – Female tail tips; I – Entire male; J – Male head end; K – Male tail.

11. HIRSCHMANNIELLA MUCRONATA (DAS, 1960) LUC & GOODEY, 1963

(Fig. 7, A-B)

Syn. *Radopholus mucronatus* Das, 1960
R. oryzae Timm, 1956
Hirschmanniella magna Siddiqi, 1966

Measurements

Females: L = 1.67-2.20 mm; a = 57-60; b = 11-14; b′ = 4.6-5.2;
c = 18-20; c′ = 4.1-5.1; V = 49-53; o = 10-16.

Males: L = 1.70-1.83 mm; a = 52-60; b = 12-14; b′ = 4.7-5.4;
c = 17-21; c′ = 4.1-5.0; o = 11-19.

Description

Female: Body slightly ventrally curved. Lip region continuous, low, hemispherical, marked by four to six annules. Lateral fields marked by four incisures, areolated in posterior region of body. Stylet robust, 24-29 μm long; with rounded basal knobs, slightly sloping anteriorly. Oesophagus typical of the genus. Female reproductive system amphidelphic. Spermatheca with sperm. Intestine slightly overlapping rectum. Tail terminus pointed, usually with a ventral projection or mucro. Phasmids in posterior third of tail.

Male: Similar to female in general shape and morphology except for sexual dimorphism and tail shape. Spicules 29-36 μm long. Gubernaculum 9-14 μm long.

Remarks: *H. mucronata* can be distinguished from *H. gracilis* by the hemispherical lip region, longer stylet (stylet 21-24 μm in *H. gracilis*), lateral fields with distinct areolations in posterior region of body.

This is a widely distributed species in the Indian states of Andhra Pradesh and Orissa and is regarded as a key or potential pest of rice crop. It is also reported as a potential pest of rice from the Philippines, Thailand and Bangladesh.

12. HIRSCHMANNIELLA SPINICAUDATA (SCH. STEK., 1944) LUC & GOODEY, 1963

(Fig. 7, C-K)

Syn. *Tylenchorhynchus spinicaudatus* Sch.Stekh., 1944
Hirschmannia spinicaudata (Sch.Stekh., 1944), Luc & Goodey, 1962
Radopholus lavaleri Luc, 1957

Measurements

Females: L = 2.10-3.54 mm; a = 42-66; b = 12-16; b′ = 4.8-8.1;
c = 19-25; c′ = 3.1-4.3; V = 51-57; o = 3-8.

Males: L = 2.33-2.45 mm; a = 52-73; b = 11-16; b′ = 4.2-6.9;
c = 19-23; c′ = 3.4-5.0; o = 3-8.

Description

Female: Body slightly curved ventrally. Lip region continuous, hemispherical or slightly conical, marked by four to five annules. Lateral fields marked by four incisures, may be areolated. Stylet robust, 40-50 μm long; with well-developed rounded basal knobs, slightly sloping posteriorly. Oesophagus typical of the genus. Female reproductive system amphidelphic. Spermatheca functional. Intestine distinctly overlapping rectum. Tail terminus pointed to rounded. Phasmids located in posterior third of tail.

Male: Similar to female in general shape and morphology except for the reproductive system and the tail shape. Spicules 41-54 μm long along the median line. Gubernaculum 13-19 μm long. Bursa subterminal.

Remarks: This is a widely distributed species in African countries, Cuba, Venezuela, Vietnam and Ivory Coast. Luc & Fortuner (1975) have redescribed this species.

13. **HIRSCHMANNIELLA CAUDACRENA** SHER, 1968

(Fig. 8, E-G)

Measurements

Females: L = 1.08-1.79 mm; a = 46-62; b = 12-16; b′ = 4.3-6.6;
c = 12-16; c′ = 4.8-7.3; V = 51-57; o = 11-12.

Males: L = 1.11-1.62 mm; a = 42-62; b = 13-17; b′ = 4.1-5.7;
c = 13-16; c′ = 5.1-6.9; o = 12-21.

Description

Female: Body slightly ventrally curved upon fixation. Lateral fields marked by four incisures, distinctly areolated in the posterior region of the body. Lip region continuous with body, flattened anteriorly, marked by four annules. Excretory pore anterior to the level of oesophago-intestinal junction. Stylet 18-20 μm long, with rounded basal knobs. Oesophagus typical of the genus. Intestine not overlapping rectum. Female reproductive system amphidelphic. Tail terminus with a pointed projection which is marked by a notch-like structure ventrally. Phasmids located in the posterior third of tail.

Male: Similar to female in general shape and morphology except for the reproductive system. Spicules 29-34 μm long. Gubernaculum 8-12 μm long. Bursa subterminal.

Remarks: This species has been reported from around roots of rice from the U.S.A., Korea, Taiwan, Hong Kong, Vietnam, New Zealand and Nigeria.

14. **HIRSCHMANNIELLA BELLI** SHER, 1968

(Fig. 8, A-D)

Measurements

Females: L = 1.61-2.22 mm; a = 58-78; b = 11-16; b′ = 5.3-7.8;
c = 15-19; c′ = 4.7-6.8; V = 50-55; o = 11-21.

Males: L = 1.42-1.90 mm; a = 59-79; b = 11.2-14.1; b′ = 5.1-6.2;
c = 15-18; c′ = 5.2-6.6; o = 14-17.

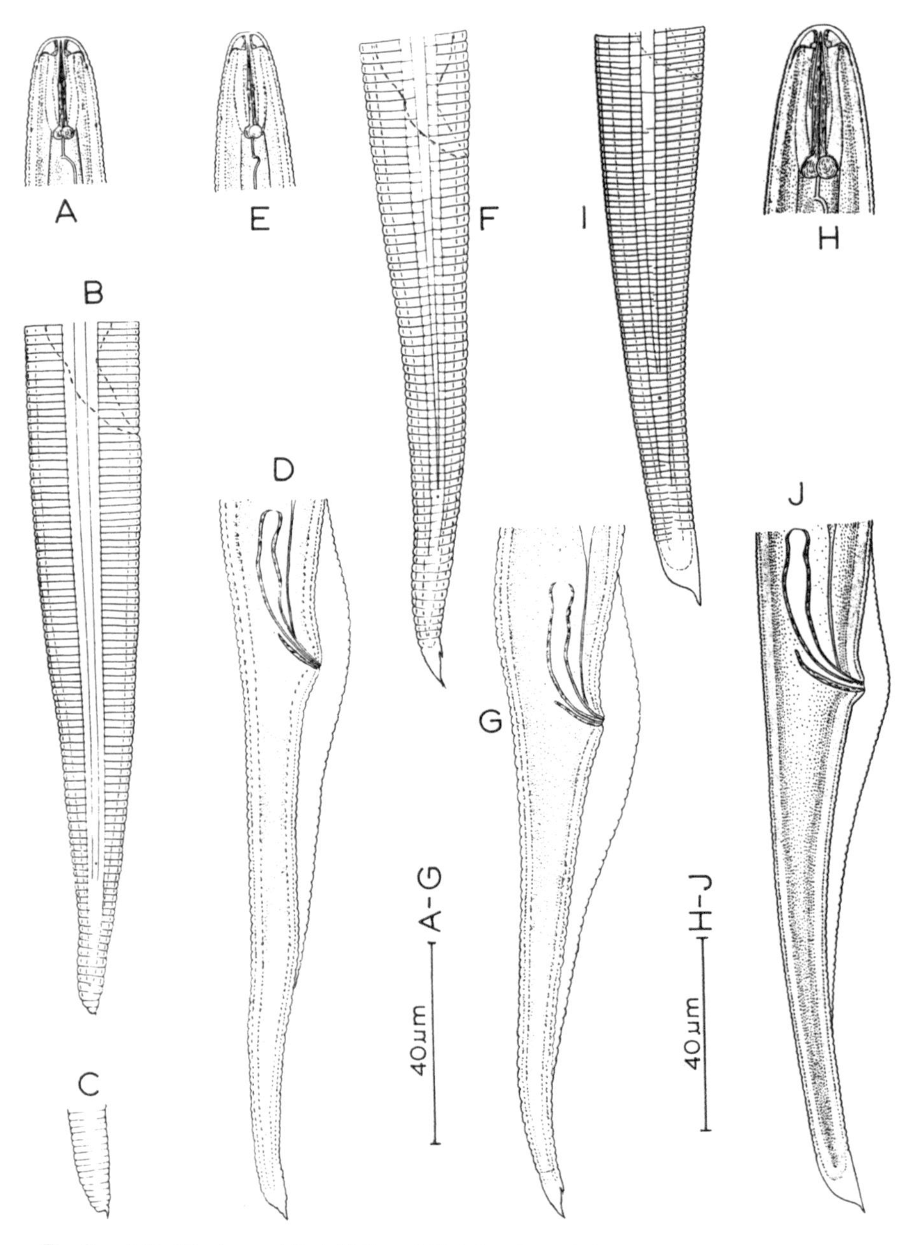

Fig. 8: A-D: *Hirschmanniella belli* Sher, 1968 (after Sher, 1968). A – Head end; B – Female tail; C – Female tail tip; D – Male tail. E-G: *Hirschmanniella caudacrena* Sher, 1968 (after Sher, 1968). E – Head end; F – Female tail; G – Male tail. H-J: *Hirschmanniella imamuri* Sher, 1968 (after Sher, 1968). H – Head end; I – Female tail; J – Male tail.

Description

Female: Body slightly ventrally curved. Lateral fields marked by four incisures, not areolated. Lip region continuous with body, marked by three distinct annules, flattened anteriorly with rounded edges. Excretory pore slightly anterior to the level of oesophago-intestinal junction. Stylet 20-22 μm long; with rounded basal knobs having sloping anterior surface. Oesophagus typical of the genus. Female reproductive system amphidelphic. Spermatheca without sperm. Intestine not overlapping rectum. Tail terminus annulated, with distinct ventral mucro.

Male: Similar to female except for the reproductive system, slightly higher lip region and tail shape. Lip region marked by four annules. Spicules 31-36 μm long. Gubernaculum 8-10 μm long. Bursa subterminal. Tail terminus with ventral projection.

Remarks: The species is commonly found in paddy fields in the U.S.A.

15. **HIRSCHMANNIELLA IMAMURI** SHER, 1968
(Fig. 8, H-J)

Measurements

Females: L = 2.15-2.72 mm; a = 62-96; b = 12-18; b′ = 5.7-7.6;
c = 17-23; c′ = 3.9-6.1; V = 48-54; o = 10-15.

Males: L = 2.18-2.61 mm; a = 67-88; b = 13-18; b′ = 5.3-8.3;
c = 17-22; c′ = 5.0-6.4; o = 10-16.

Description

Female: Body slightly ventrally curved upon fixation. Lateral fields marked by four incisures; completely areolated in the posterior region of body. Lip region continuous, hemispherical, marked by six (five to seven) annules. Stylet robust, 29-32 μm long, with well-developed rounded basal knobs. Excretory pore slightly anterior to oesophago-intestinal junction. Tail cylindrical, terminus with a pointed ventral digitate process. Cuticle becoming smooth much above the tail terminus. Phasmids located in the posterior third of tail region.

Male: Similar to female in general shape and morphology except for the reproductive system and tail. Spicules 37-44 μm long. Gubernaculum 10-14 μm long. Bursa subterminal.

Remarks: *H. imamuri* has been reported as a potential pest of rice from Japan, Korea, Nigeria and Vietnam.

RICE ROOT NEMATODES
Symptoms, Loss and Control

Since most of the *Hirschmanniella* species are associated with rice, the species of this genus are commonly called "rice root nematodes". The host and parasite relationship is so strong that rice root nematodes have been found in all rice growing countries. In the absence of rice, the weeds, cabbage and other vegetables, cotton, maize and sugarcane etc. may serve as hosts for *Hirschmanniella* species. Their population is generally higher in the damp paddy soils than in well-drained fields because they are well adapted to survive in saturated moist soils. However, they can also survive in extreme conditions, that is, under severe dry conditions up to five months as well as under flooded conditions for at least seven months.

As soon as the rice seeds germinate or the seedlings are transplanted they penetrate at any point of roots, from tip to the base. All the stages can penetrate and move freely inside the roots. They feed mostly on the cells around the base of the root hairs.

Symptoms

There is no specific disease caused by rice root nematodes and as such there are no specific above ground symptoms. However, due to damage to roots, retardation of growth, stunting of plants, chlorosis and the reduction in the number of tillers may be observed when compared with the normal plants. These symptoms cannot be considered as diagnostic because similar symptoms may also be caused by other factors in the above-ground parts. In case of severe infestations, the late maturation of crop can also be observed. Sometimes the injured roots may be invaded by bacteria and fungi (secondary infections). Under these conditions, the roots become brown then black and finally rot. As a result, the whole root system of the infected plants are much reduced and more darker than the uninfected plants. If a section of infected roots is cut, necrosis and formation of cavities due to disruption of cell walls can be observed.

Yield Losses

The loss in rice yield is estimated to be 10-36% in Asian and African countries. Ahmad et al. (1984) have estimated the yield loss to be 12.5-19% due to *Hirschmanniella gracilis*. In general, an average of 25% yield loss has been found in rice due to *Hirschmanniella* species in India (Panda & Rao, 1969; Rao & Panda, 1970).

Life Cycle

Most *Hirschmanniella* species are bisexual. The eggs are laid in the roots. The juveniles hatch and develop in the root cortex. Since hatching and developmental stages may take place within the host plant, the number of nematodes increases with the development of root system. The different juvenile stages may be identified under stereoscopic binocular microscope on the basis of body, developing gonad and stylet lengths. The nematode population gradually declines in roots when the panicles appear or more so when the root system starts degenerating near the time of harvesting (Das et al.,

1984). There may be one or two generations per crop. This may be ascertained on the presence of gravid females in the root population.

Control

Considerable work has been done on the varieties of rice resistant against rice root nematodes but with limited success. Fairly resistant varieties have been found containing one nematode per gm as against 100 nematodes per gm root in susceptible varieties. Since *Hirshmanniella* species have an extensive host range on grasses and sedges, weed control brings down their population significantly in the fields.

The application of straw ashes, neem and mustard cakes reduces the rice root nematode populations in the soil as well as roots. Baqri et al. (1988) have also found that incorporation of fresh chopped leaves of water hyacinth @ 19 kg/16 sq.ft plot at the time of main field preparation before seedling transplantation brings down the soil and root populations of *Hirschmanniella gracilis,* that is, 36-56% reduction in soil while 36-62% in roots. This reduction resulted in 27-29% increase in the yield over the untreated plots. Intercropping with trap plants like *Sesbania rostrata* or *Sphenoclea zeylanica* also helps in reducing the rice root nematode populations.

Ahmad et al. (1984) have reported from West Bengal that the soil treatment with carbofuran granules @ 1 kg a.i./ha in the seed bed and in the main field a day before seed sowing or seedling transplantation brings down the *Hirschmanniella gracilis* population significantly in rice roots as well as in the field. The main field should also be treated the second time with the same dose of carbofuran granules on the fiftieth day after transplanting. It has been noted that the grain yield increases by 12.5-19% in the carbofuran treated subplots over the untreated control.

The soil treatment with DBCP, DD, metham sodium and phosphamidon etc. also reduces the rice root nematode population and increases the rice yield. The root dip treatment of seedlings for 12 hours in the above-mentioned nematicides (0.2% solution), before transplanting, has also been recommended.

16. DITYLENCHUS ANGUSTUS (BUTLER, 1913) FILIPJEV, 1934 (Fig. 9)

Syn. *Tylenchus angustus* Butler, 1913
Anguillulina angustus (Butler) Goodey, 1932

Measurements

Females: L = 0.8-1.20 mm; a = 50-62; b = 6-9; c = 18-24; V = 78-80.
Males: L = 0.70-1.18 mm; a = 40-55; b = 6-8; c = 19-26; T = 60.

Description

Female: Body almost straight or slightly ventrally curved upon fixation. Cuticle transversely striated, 1 μm apart at mid-body. Lip region smooth,

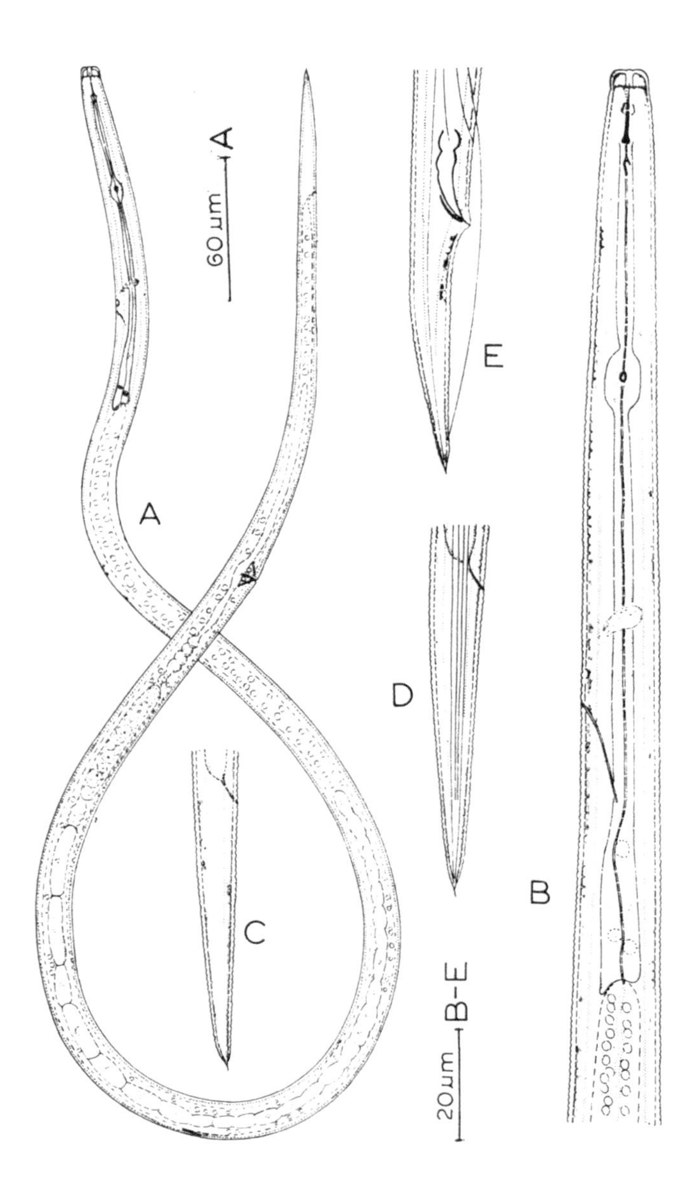

Fig. 9: *Ditylenchus angustus* (Butler, 1913) Filipjev, 1934 (after Seshadri & Dasgupta, 1975). A – Entire female; B – Oesophageal region; C & D – Female tails; E – Male tail.

slightly marked off from body, low, wider than height at lip base, apex flattened. Cephalic framework very lightly sclerotized. Lateral fields marked with four incisures, outer ones more distinct. Stylet 9-11 μm long, conus about 45% of stylet length with posteriorly sloping basal knobs. Median oesophageal bulb oval, with a distinct valvular apparatus; posterior oesophageal bulb usually clavate, slightly overlapping the intestine, mostly ventrally, with three distinct gland nuclei. Cardia absent. Female reproductive system prodelphic. Vagina somewhat oblique. Spermatheca elongated, filled with rounded sperm. Tail 5.2-5.4 anal body-widths long, tapering to a sharply pointed terminus resembling mucro. Phasmids just behind middle of tail, small pore-like and difficult to see.

Male: Abundant. Similar to female in general shape and morphology except for reproductive system and tail shape. Bursa (caudal alae) almost terminal. Spicules curved ventrally, 16-21 μm long. Gubernaculum 6-9 μm long.

Larvae: Similar to adults. Oesophagus proportionally longer than adults.

Remarks: *Ditylenchus angustus* is a serious pest of rice, causing 'ufra' disease. Seshadri and Dasgupta (1975) have provided a detailed redescription of this species.

STEM NEMATODE : UFRA

Ditylenchus angustus is an obligate and serious pest of rice causing the well-known 'ufra' disease. In addition, this may also attack duck weed *(Hygrophyz aristata)* and swamp rice grass *(Leersia hexandra)*. Ufra was first reported by Butler (1913a) from the then East Bengal in India, now Bangladesh. The local meaning of ufra is "burnt by lightning" or "gone away". In Bangladesh the disease is also called '*dakpora*'. In Thailand the people call the disease "*Yed Ngo*" which means "twisty disease". Since *D. angustus* spreads usually through flood waters, ufra is mainly found in deep-water paddy. However, it has also been reported from the irrigated fields adjacent to deep-water fields (Bakr, 1977; Miah & Rahman, 1981; Miah, 1984). The disease has been recorded from Bangladesh, Burma, Egypt, India (Uttar Pradesh, Assam, Bihar and West Bengal), Madagascar, Malaysia, Philippines, South Thailand and Vietnam.

Ditylenchus angustus is generally called 'stem nematode' because of its stem inhabiting nature. They feed ecto-parasitically on younger or soft leaves, leaf sheaths, pedencles and spikelets. After harvesting of infested crop, the nematodes can survive in the fallow fields and wait for the next rice crop on wild rice, stubbles left in the field or the weed grass. They live in the coiled condition (inactive) on the dried parts of plants left in the fields. As soon as the rice seeds germinate or the seedlings are transplanted in the infested fields, the nematodes invade the plants and appear on the terminal buds. Since they prefer to feed on younger tissues, they shift accordingly with the growth of plants towards the growing tips. During flowering they are either found on the stem just above the nodes or between the leaf sheaths or peduncles and suck cell contents continuously through stylet. Hence, all the damage to plants or crops is due to continual removal of the cell sap. The nematodes move towards the young leaves through the space in between the leaf sheath. At four leaf stage of rice plant, it usually takes about three days to reach the inner most leaf sheath. In rice plants, the inner most leaf-fold is generally so compact that it checks the movement of nematode towards the top-most growing point of plant. Perhaps this is the reason that the infested plants are not killed and can still grow (Butler, 1913b).

While studying the population dynamics of *D. angustus*, Cox and Rahman (1979a) have observed three peaks during the crop period. In the first peak, the number of nematodes was more than 3000 per stem. In the second peak it was 1000 to 2000 per stem while in the third peak the number was only about 1000 per stem. The first peak represents primary infestation, second the secondary and the third peak the tertiary indicating development of Ufra I, II and III respectively.

Symptoms

It is generally difficult to detect the symptoms of ufra in the seedlings or in the early age of the crop up to eight weeks. Sometimes chlorosis may be observed in the infested young seedlings within the week after infection. The young leaves may be slightly marked by pale longitudinal streak and the growth of plants may be somewhat stunted. The leaves may be slightly thinner and more flaccid than the normal.

The symptoms of ufra are well established in two months old crops. The chlorotic portion of leaf becomes brown to dark brown. Later, the scattered dark stains also appear on the upper internodes of the stem which become completely dark brown. As the age of plants advances, the disease also advances. The twisting of leaf and leaf sheath are the commonly found symptoms. In more severe and advanced stage of disease, the leaf margins become corrugated. Sometimes many branches grow from the infested node which gives a bushy appearance to the plants. The panicles may remain enclosed within the flag leaf sheath or may emerge partially or fully from it. Finally the fertility and grain filling declines and malformation of grains takes place. The grains may become brownish or remain unfilled (Plate 1).

On the basis of emergence, non-emergence or partial emergence of panicles, Butler (1913b) has classified the disease in two categories which are called 'ufra I' and 'ufra II'.

UFRA I – This has been referred to as *thor ufra* or swollen ufra. In this category the panicles do not emerge and are completely enclosed within the flag leaf sheath. These symptoms occur due to severe infection of *D. angustus*

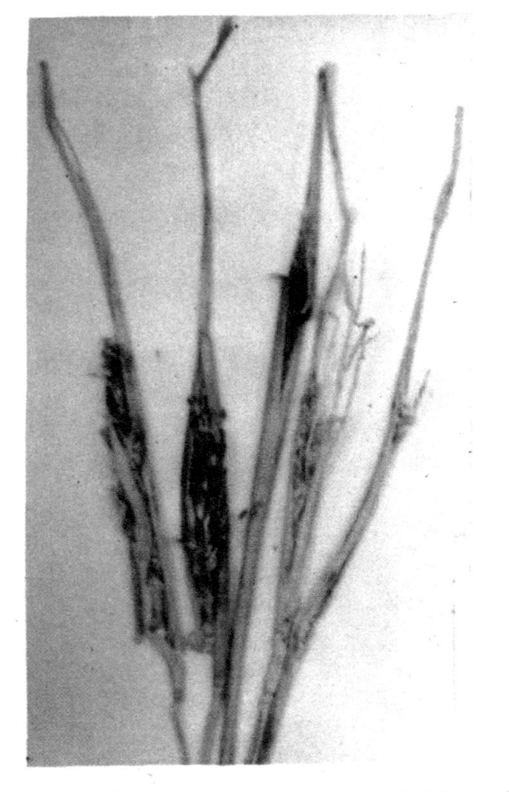

Plate 1: Symptoms of ufra disease (Courtesy M.F. Rahman, Deptt. of Nematology, Assam Agricultural University)

which damage the panicles at early stage. Sometimes branching of the stem can also be observed.

UFRA II – This category is equivalent to *pucca ufra* or ripe ufra. There is a partial emergence of panicles because the damage is done at a later stage. The panicles may bear unfilled grains on the lower part.

These two types of ufra have been expanded into three by Cox & Rahman (1980). They proposed the term 'UFRA III' for the intermediate type where the panicles emerge completely but with many unfilled or empty grains.

Ufra I and Ufra II show most severe symptoms in which the yield may decline up to zero level while in Ufra III a little yield (10-30%) may be expected.

In most fields ufra symptoms appear in patches which are generally enlarged as the crop age advances. It has been observed that the centre of each patch is more infested than the plants at periphery. In some fields the symptoms are uniformly distributed. Perhaps this is determined by the movement of water.

Life History and Disease Cycle

Left over stubbles, ratoons, wild rice and some weeds in the field are the primary source of *D. angustus* infection. As the crop approaches maturity, the nematodes (adult and fourth stage juveniles) cease feeding and coil up in dried parts of rice plants. These nematodes may survive for six to 15 months (Butler, 1913a; Miah & Bakr, 1977). On the availability of rain, tidal or flood waters, the nematodes on the left-over stubbles or in the empty grains etc. become uncoiled and active within an hour in search of the host plants. The nematodes can survive in water up to four months. As the seeds germinate or the seedlings are transplanted, the nematodes start feeding on them. The reproduction takes place inside the host plants during May-June and November, usually after tillering stage. Butler (1913b) has recorded three generations in one season. Each female can produce 50 to 100 eggs (Butler, 1913a). Miah & Bakr (1977) have observed that 28°-30°C temperature and more than 80% relative humidity are favourable factors for infection, disease development and reproduction. The first moulting takes place within the egg shells. The developmental cycle from the second stage larva takes only 15 days and the whole cycle is completed within 24 days in the artificial conditions. Under normal conditions (air temperature, relative humidity and availability of the host), the duration of life cycle is likely to be shorter because the eggs are directly laid on the rice stem and the second stage juveniles do not waste time searching for the host plants. All the four juvenile stages may be recognized mainly on the basis of differences in body lengths.

The population of adults reaches peak levels in August while the plants are still young under the peak flood depth. The number of nematodes varying one to 3000 per plant. Catling et al. (1979) have recorded 30,000 nematodes/stem. Adults and juveniles of all stages are found on the plants.

Yield Losses

It is estimated that 4 million ha rice crop is infested with *D. angustus*. The yield loss generally depends on the source, inoculum density, time of infection and the variety grown in the field.

Butler (1913b) has estimated the loss of grain to be nearly 8,000 tons in Noakhali area of Bangladesh. The yield loss increases if infection takes place at the seedling stage of the crop because the young plants are more susceptible than the older plants. Mondal et al. (1989) have observed that only 4-10% infested seedlings at the time of transplanting are enough to cause severe damage. Rahman & Evans (1987) have noticed that the highest infestation (73%) occurs when the nematode infested material was mixed with soil at sowing time. This loss comes down to 60% and 13% when the seedlings are inoculated at four weeks and six weeks respectively after sowing. The severity of ufra also increases if the monsoon is early and wet.

The overall loss due to ufra has been estimated from 20-90% with an average loss at 30%. Catling et al. (1978) have recorded 20% loss annually of the total deep-water rice crop in Bangladesh. On individual field basis, Miah & Bakr (1977) have estimated 40-60% or occasionally 100% loss. If there are more than 40% ufra II symptoms, the yield loss may be up to 100% (Cox & Rahman 1980). Singh (1953) has reported 5-50% loss in ufra infested fields in India while Hashioka (1963) has estimated 20-90% loss in Thailand.

Association with other Pathogens and Zinc Deficiency

The symptoms of blast fungus (*Pyricularia oryzae*), sheath rot fungus (*Sarocladium oryzae*) and bacterial leaf blight *Xanthomonas compestris* are commonly found in ufra infested fields. Mondal et al. (1986) have observed that the combined effect of *D. angustus* and blast fungus increases severity in ufra disease. Mondal et al. (1985) have also noted that the combined effect of ufra and sheath rot increases the yield loss compared to their individual effect. Similar results have been found when ufra is combined with bacterial leaf blight. Association of leaf and node blast on ufra infested stem has also been reported by Rathaiah (1988).

Zinc deficiency in *D. angustus* infested rice fields and pots has been causing more damage to the crop (Miah et al., 1984). Sometimes, the high nitrogen contents in ufra infested plants may increase the blast susceptibility which causes more severity due to the combined effect of both (Mondal et al., 1985 & 1986).

Control

The population of *D. angustus* can be brought down through cultural and management practices, crop rotation and chemicals.

Since the stem nematodes generally wait on stubbles, ratoons, wild rice and some weeds for the next crop, the burning of all these left overs after harvesting is a very effective method for the control of ufra disease. The left over stubbles etc. should be uprooted and allowed to be dried in the field before they are set on fire. The fields may be ploughed after harvesting so that the stubbles are decomposed and the nematodes are exposed to the sun for a long period. As a result it will be difficult for the nematodes to survive till the next crop. Many farmers cut leaves and the upper portion of the stems showing ufra symptoms. This old practice also helps in reducing the nematode population from the field.

Crop rotation is also an effective method for the management of *D. angustus* population. Rice, if rotated with mustard reduces the population of *D. angustus* considerably. Growing jute in ufra infested fields also helps in the reduction of ufra severity. If the sowing or transplanting of deep-water rice is delayed by two to three weeks than the normal time or till the first floods are over then ufra infestation is reduced by 25-28% or more (McGeachie & Rahman, 1983).

Some varieties of deep-water rice and wild rice have also been reported as resistant to ufra disease. Recently, more than 3000 varieties from India and Bangladesh have been screened. Out of these, only 116 (CNL of India, Rayada, Brazil and Karkate etc.) have been found to be considerably resistant.

Though the chemical control is effective, it is costly, risky and hazardous. Moreover, the use of chemicals in deep-water rice is very limited because of difficulties in their application and it may become more hazardous if applied during floods. However, a good number of chemicals have been found very effective against ufra disease.

Rahman & Miah (1989) have observed that the application of carbofuran 3G @ 1.0 kg a.i./ha just before broadcasting or transplanting brings down the ufra infestation by 42-63% with an increase in the yield by 38-59% over the untreated plots. In ufra infested fields having 0.5 m flood water, the application of carbofuran 3G @ 24 kg a.i./ha followed by a foliar spray of benlate 50 wp (benomyle) @ 2.5 kg a.i./ha significantly reduces ufra and increases the yield (Cox & Rahman, 1979b). The soil treatment with diazinon, disulfoton and fensulfothion is also effective. In zinc deficient fields, the application of zinc also reduces the severity of disease. Two spray treatments of hexadrin (1:700 in water) before preflowering stage have been reported significantly effective.

Root dip treatment of seedlings in nematicides solution is less expensive and least hazardous. The solutions of Miral 3% and Tecto 40 FL at 7.5-10% and 2.5-10% respectively have been suggested to control ufra disease through root dip treatment of seedlings. The roots of ufra infested seedlings may be soaked for 18 hours in these solutions before transplanting.

Attempts have also been made to control ufra disease through root dip treatment of seedlings in leaf extracts of *Acacia* sp., neem *(Azadirachta indica)* and Ata (*Anona* sp.) etc. None of the plants evaluated for root dip treatment of seedlings in leaf extractions has been found to be significantly effective.

17. **MELOIDOGYNE JAVANICA** (TREUB, 1885) CHITWOOD, 1949

(Fig. 10)

Syn. *Heterodera javanica* Trueb, 1885
Tylenchus (Heterodera) javanica (Treub) Cobb, 1890
Anguillula javanica (Treub) Laverang, 1901
Meloidogyne javanica javanica Gillard & van den Brande, 1955
Meloidogyne javanica bauruensis Lordello, 1956

Measurements

Females: L = 0.541-0.804 mm; width = 311-581 μm; stylet = 14-18 μm; stylet knobs = 2-5 μm; dorsal oesophageal gland opening = 2-5 μm behind stylet.

Males: L = 0.757-1.297 mm; a = 17.5-4.3; stylet = 20-23; stylet knobs = 3.6-5.4 μm; spicules = 21-32 μm; gubernaculum = 7.2-9.4 μm.

Second stage juveniles: L = 0.387-0.459 mm; a = 27-36; b = 2.1-3.3; b′ = 7.1-8.0; stylet = 9.4-11.4 μm.

Description

Female: Body almost spherical with projecting neck, rounded posteriorly but perineal region protruding slightly. Lip region slightly wider than adjoining body, marked by one annule behind head cap. Spear slender, dorsally curved, with rounded basal knobs. Perineal pattern (= posterior cuticular pattern) round or oval to pear-shaped, dorsal arch varying from rounded, to moderate height, may be flattened dorsally. Striae smooth to wavy. Lateral fields marked by two incisures, dividing pattern into dorsal and ventral sectors.

Male: Lip region rounded in dorso-ventral view, continuous with body; the basal annule wider than first annule, the former giving the appearance of one annule on one side and two on the other side. Amphidial apertures distinct. Stylet knobs rounded. Intersexes common. Lateral fields marked with four incisures, the outer bands may be areolated. Tail shape variable, digitate in lateral view and bluntly rounded in ventral. Phasmids usually at cloacal level. Spicules slightly curved. Gubernaculum thin, crescentic.

Second stage juvenile: Lip region continuous, truncate cone-shaped in lateral view, marked by three annules behind head cap. Stylet knobs not prominent. Lateral fields marked by four incisures. Tail tapering to subacute or finely rounded terminus.

Remarks: *M. javanica* also has a wide range of hosts and distribution in warm and tropical regions of the world. Orton (1972) has provided a detailed redescription of this species.

18. **MELOIDOGYNE INCOGNITA** (KOFOID & WHITE, 1919) CHITWOOD, 1949

(Fig. 11)

Syn. *Oxyuris incognita* Kofoid & White, 1919

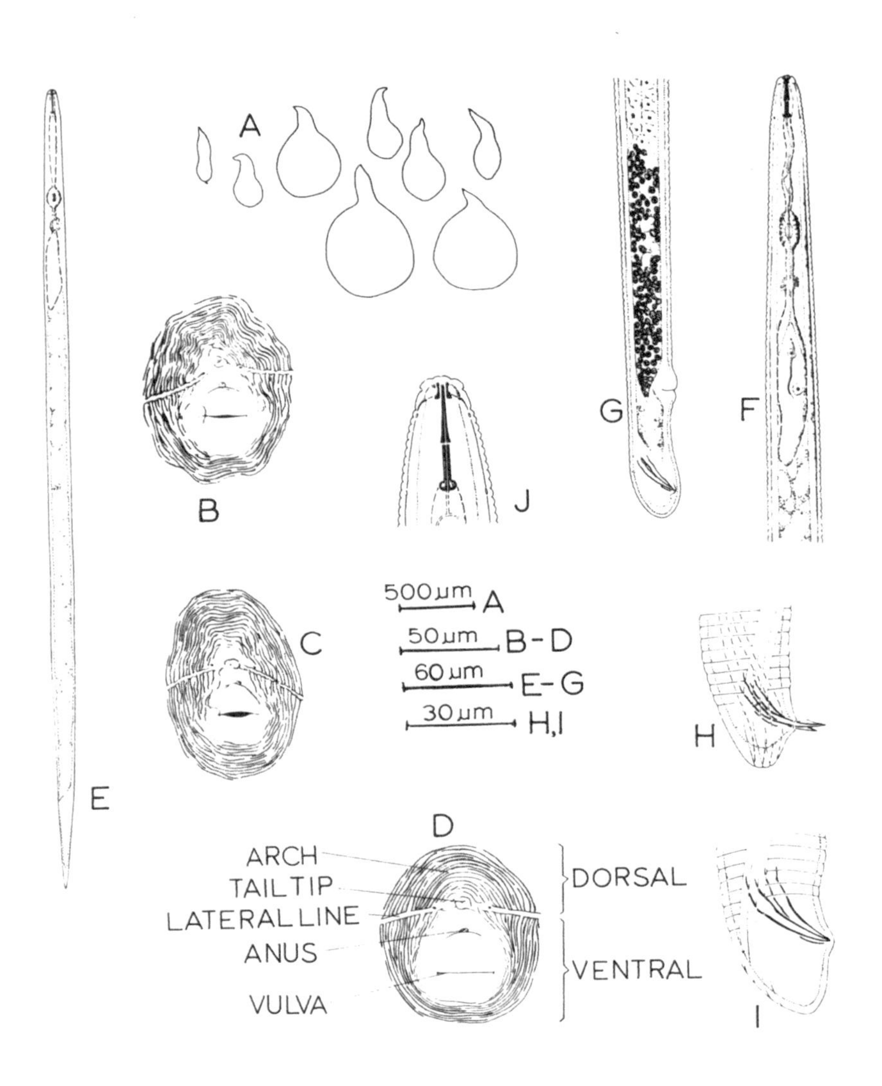

Fig. 10: *Meloidogyne javanica* (Treub, 1885) Chitwood, 1949 (after Orton, 1972). A-D: Female. A – Shapes of immature and adult females; B-D – Posterior cuticular patterns. E – Second stage juvenile. F-I: Male. F – Oesophageal region; G – Posterior region; H & I – Tails.

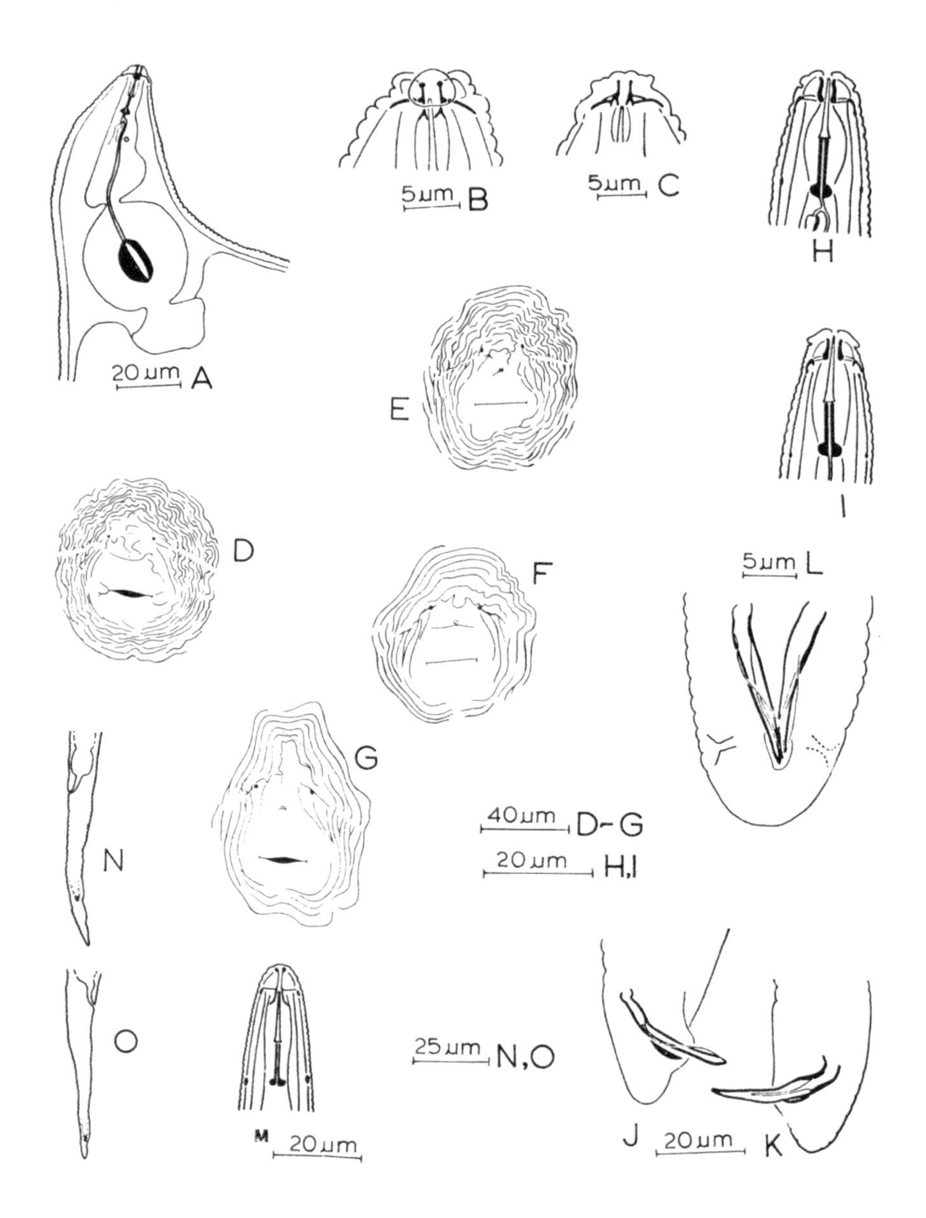

Fig. 11: *Meloidogyne incognita* (Kofoid & White, 1919) Chitwood, 1949 (after Orton, 1975). A-G: Female. A – Anterior end; B & C – Head ends; D-G – Posterior cuticular patterns. H-L: Males. H & I – Head ends; J & K – Tails (lateral view); L – Tail (ventral view). M-O: Second stage juvenile. M – Head end; N & O – Tails.

Measurements

Females: L = 0.500-0.723 mm; width = 331-520 μm; stylet = 13-16 μm; width of stylet base = 3-5 μm; orifice of dorsal oesophageal gland = 2-4 μm.

Males: L = 1.108-1.953 mm; a = 31-35; head height = 6.8-8.6 μm; stylet = 23-32.7 μm; width of stylet base = 4.7-6.8; orifice of dorsal oesophageal gland = 1.4- 2.5 μm; spicules = 28.8-40.3 μm; gubernaculum = 9.4-13.7 μm.

Second stage juveniles: L = 0.337-0.403 mm; a = 24.9-31.5; b = 2.02-2.14; b′ = 6.4-8.4; tail length = 38-55 μm; stylet = 9.6-11.7 μm.

Description

Female: Body spherical with projecting neck. Lip region bearing two or three annules behind head cap. Cuticle thickens abruptly at base of stylet. Stylet knobs rounded. Excretory pore 10 to 20 annules behind lip region. Posterior cuticular pattern highly variable, typical "incognita type" with striae closely spaced very wavy to zig-zag, specially dorsally and laterally. Lateral fields not clear, sometimes marked by breaks in striae, broken ends often forked, "acreta type" with striae smoother, more widely spaced. Dorsal arch variable. Striae often forked along a 'lateral line'.

Male: Body showing marked sexual dimorphism, slender. Cuticle transversely striated. Lateral fields marked by four incisures, outer bands areolated, inner bands may be cross striated. Lip region almost continuous with body, high, truncate cone-shaped, head cap with stepped outline in lateral view. Lip region marked by distinct annules, one to three on sublateral head sectors and one to five on lateral head sectors. Stylet conus longer than shaft, knobs usually wider than length with flat concave or 'toothed' anterior margins. Excretory pore posterior to isthmus. Testes one or two. Tail bluntly rounded with smooth terminus. Phasmids nearly at cloacal level. Spicules slightly curved. Gubernaculum crescentric. Bursa absent.

Second stage juveniles: Lateral fields with four incisures. Lip region almost continuous, truncate cone-shaped in lateral view, subspherical in dorso-ventral view. Head cap wide. Lip region marked by two annules in lateral view and three annules in dorsal or ventral view. Tail tapering to subacute terminus.

Remarks: *M. incognita* has a wide range of hosts (over 700 host species and varieties) including rice throughout tropical and warmer countries. Orton (1973) has provided a detailed redescription of this species.

19. **MELOIDOGYNE GRAMINICOLA** GOLDEN & BIRCHFIELD, 1965
(Fig. 12)

Measurements

Females: L = 0.445-0.765 mm; width = 275-520 μm; a = 1.2-1.8; stylet = 10.6-11.2 μm.

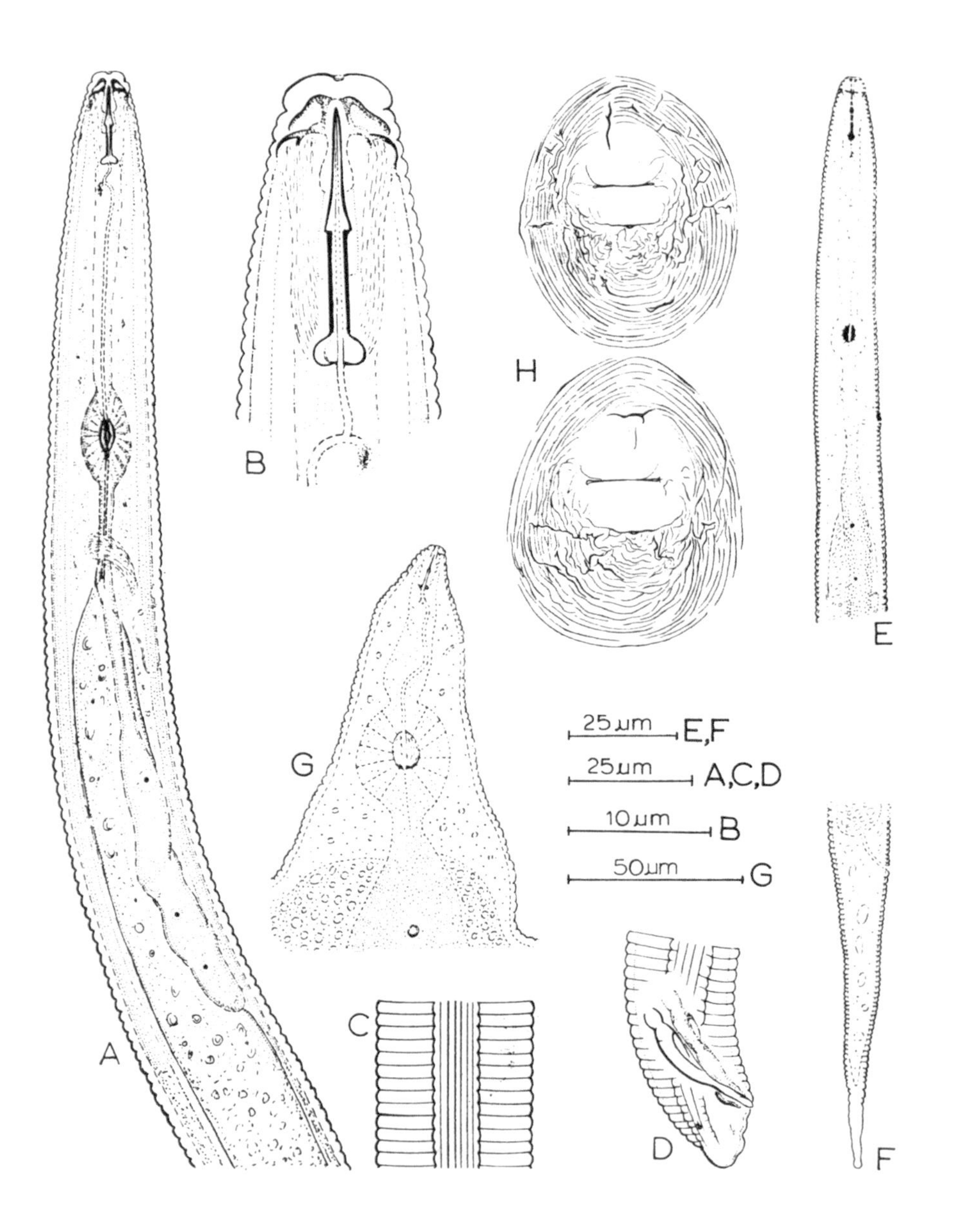

Fig. 12: *Meloidogyne graminicola* Golden & Birchfield, 1965 (after Mulk, 1976). A-D: Male. A – Oesophageal region; B – Head end; C – Lateral fields near mid-body; D – Tail. E & F: Second stage juvenile. E – Anterior region; F – Tail. G & H: Female. G – Anterior end of mature female; H – Posterior cuticular patterns.

Males: L = 1.22-1.42 mm; width = 24-35 μm; a = 72-215; stylet = 16-17.3 μm; spicules = 27-29 μm; gubernaculum = 5.6-6.7 μm.
Second stage juveniles: L = 0.415-0.844 mm; a = 22-27; b = 2.9-4.0; c = 5.5-6.7; stylet = 11.2-12.3 μm.

Description

Female: Body pearly white, globular to pear-shaped and with a small neck. Cuticle striated, sometimes with fine irregular punctations. Lip region slightly marked off from body, smooth, cephalic framework weak. Stylet small and delicate, with rounded basal knobs sloping posteriorly. Orifice of dorsal oesophageal gland 3-4 μm from stylet base. Oesophago-intestinal junction obscure. Excretory pore about 1.5 stylet length from basal knobs. Rectal glands six, prominent. Vulva and anus terminal. Perineal pattern prominent with characteristic striations. Eggs deposited in gelatinous matrix.

Male: Body vermiform. Cuticle distinctly striated, about 2 μm apart. Lateral fields marked by usually four incisures in young and eight incisures in old specimens; outer lines crenate with outer bands often areolated. Stylet stout with rounded knobs. Testis one. Spicules arcuate. Bursa absent. Tail 6-15 μm long. Phasmids small, located at about mid-tail.

Second stage juveniles: Body slender, cuticle finely striated, 1 μm apart; lateral fields marked by four incisures. Lip region almost continuous with body, marked by three post-labial annules. Cephalic framework weak. Stylet delicate with rounded knobs sloping posteriorly. Oesophagus typical of *Meloidogyne* second stage juveniles. Tail 67-76 μm long with 14-21 μm long hyaline terminal part, elongate-conoid, with rounded or clavate terminus; four to five anal body widths long.

Remarks: Mulk (1976) has redescribed the species in detail. Baqri et al. (1991) and Baqri & Das (1991) have reported this species as a potential pest of rice in West Bengal.

ROOT KNOT NEMATODES

Symptoms, Loss and Control

The *Meloidogyne* or root-knot species are obligate endoparasites having a wide range of hosts including rice. *Meloidogyne arenaria, M. incognita* and *M. javanica* are the major pests of rice in temperate, warm and tropical areas. The distribution of *M. oryzae* and *M. solasi* is restricted to South America (Surinam, Costa Rica and Panama). *Meloidogyne graminicola* is the most widely distributed serious nematode pest of rice in India and has been considered economically important next to *Hirschmanniella* spp. It has been reported from Kerala, Madhya Pradesh, Orissa, Assam, West Bengal and Tripura. In addition this has also been reported as an important pest of rice from Bangladesh, Laos, Thailand, the U.S.A. and Vietnam. About 2,000 ha of rice fields are infested with *M. graminicola* in Northern Thailand where rice is rotated with vegetables and tobacco. The root-knot infestation of rice is mainly found in upland regions because these nematodes are poorly adapted to flooded conditions.

The commonly found species in India, *M. graminicola* does not produce galls on rice roots growing in flooded soils but may survive within roots for about five weeks. However, recently Rao (1985) has reported some damage in semi-deep water rice also. The already infected plants are unable to grow above water level. In flooded soil they may survive up to five months. The coarse textured soil is more favoured than clay soil for a high build-up of root-knot population. In the absence of rice crop, *M. graminicola* infests several species of weeds or other crops in the field.

The second stage juveniles of *M. graminicola* invade roots of young plants and giant cells are formed due to the enlargement (hypertrophy) of root tissues and may be observed within four days of invasion.

Symptoms

Chlorosis, wilting, delay in flowering of crop by 10 to 15 days, reduction in growth and number of tillers are the symptoms which may be noted in the above-ground parts of infected rice plants (Plate 2). *M. graminicola* incites five to eight giant cells causing swelling at stele. In the roots the hypertrophy and hyperplasia in meristem, cortex, endodermis and xylem cause gall formation. The galls can be seen at the ends of roots and may contain many nematodes. Ultimately the growth of the roots is checked. The occurrence of galls in the roots is the main symptom by which the presence of root-knot nematode can be ascertained (Plate 3).

Yield Loss

An average yield loss due to *M. graminicola* has been estimated at 10-20% in upland rice. In severe cases the loss may go up to 50%. In flooded rice the losses due to root-knot nematodes have not been estimated. However, loss due to *M. graminicola* in deep-water rice is inevitable because already infected plants are unable to grow above the water level. In India, the losses in grain yield have been estimated to be 16-32% (Biswas & Rao, 1971; Rao & Biswas, 1973).

Life Cycle

The females lay eggs in the root tissues. The juveniles hatch inside the galls and the second stage juveniles reinfect the same roots. All the developmental stages are found within the roots. *M. graminicola* completes its life cycle

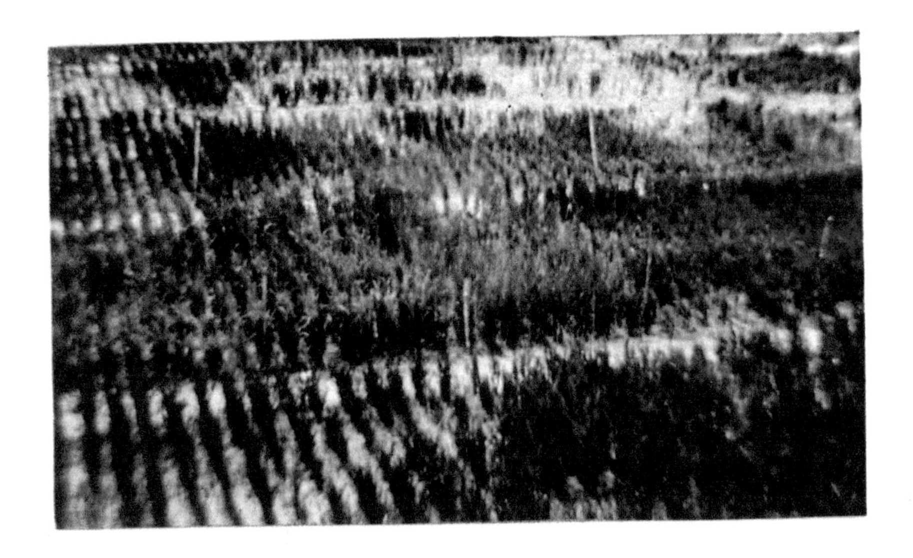

Plate 2: Rice fields showing root-knot disease symptoms (Courtesy M.F. Rahman, Department of Nematology, Assam Agricultural University)

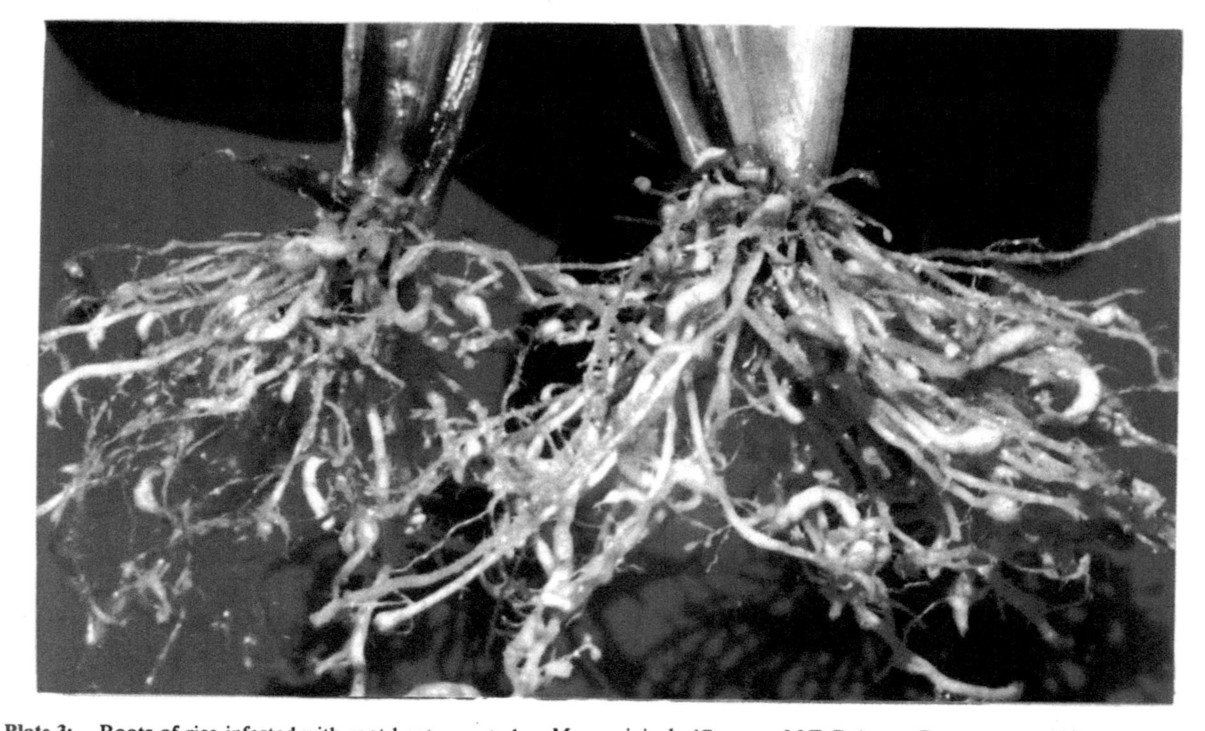

Plate 3: Roots of rice infested with root-knot nematode, *M. graminicola* (Courtesy M.F. Rahman, Department of Nematology, Assam Agricultural University)

from adult to adult stage in 19 days at 22°-29°C temperature in upland rice fields. The excess of nitrogen and phosphorus increases the progeny of *M. graminicola*. After crop harvestation the nematodes may survive in egg stage within the soil or they may continue their reproduction on other available crops or on weeds.

Control

Crop rotation with non-host crop and use of resistant varieties have been recommended as simple and effective methods for the reduction of root-knot nematode populations in the rice fields. Groundnut, sweet potato, maize, soybean and many other crops have been confirmed as non-host crops for *M. graminicola*. Since the root-knot nematodes are poorly adapted to flooded conditions, flooding the fields by irrigation water for more than 30 days also brings down the *Meloidogyne* population and increases the yield significantly. The mixed sowing of deep-water rice with sesame followed by sowing of mustard helps in reducing the nematode populations in soil. The application of rice husk, rice polish, tea waste, neem leaf dust, wheat bran and mustard cake in pots and miniplots can reduce the root-knot population by 21-40% in soil, but does not increase the yield significantly.

The soil treatment of nursery beds and the main field by carbofuran, DD, methylbromide, oxamyl is very effective for the control of these nematodes. Since the main field treatment is more costly and hazardous, root dip treatment of seedlings by these chemicals before transplanting them in the main field is also being recommended.

Jena & Rao (1973), Rao (1978, 1985) and others have reported a good number of rice varieties as resistant against *Meloidogyne graminicola*. A few of these are being mentioned here: CR 143-2-2; CR 147-2-1; CR 1009 (Sairtri); CT 428; Sudha; Mutri; Amla; Cultur 377.

20. HETERODERA ORYZICOLA RAO & JAYAPRAKASH, 1978
(Fig. 13)

Measurements

Females: L = 0.414-0.520 mm; breadth = 220-348 μm; Length/breadth = 1.5-1.9.

Cyst: Length (excluding neck) = 0.434 mm; breadth = 340 μm; length/breadth = 1.3; Eggs = 95-109 x 39-47 μm.

Males: L = 0.89-0.98 mm; width = 28 μm; a = 32-35; b = 7-9; b′ = 6-8; c = 224-326.

Second stage larvae: L = 0.37-0.42 mm; width = 18-22 μm; a = 19-20; b = 4; b′ = 2.7; c = 7.1.

Description

Female: Body lemon to broad lemon-shaped, white, asymmetrical with inclined neck. Lip region offset, marked by two annules, the second one larger.

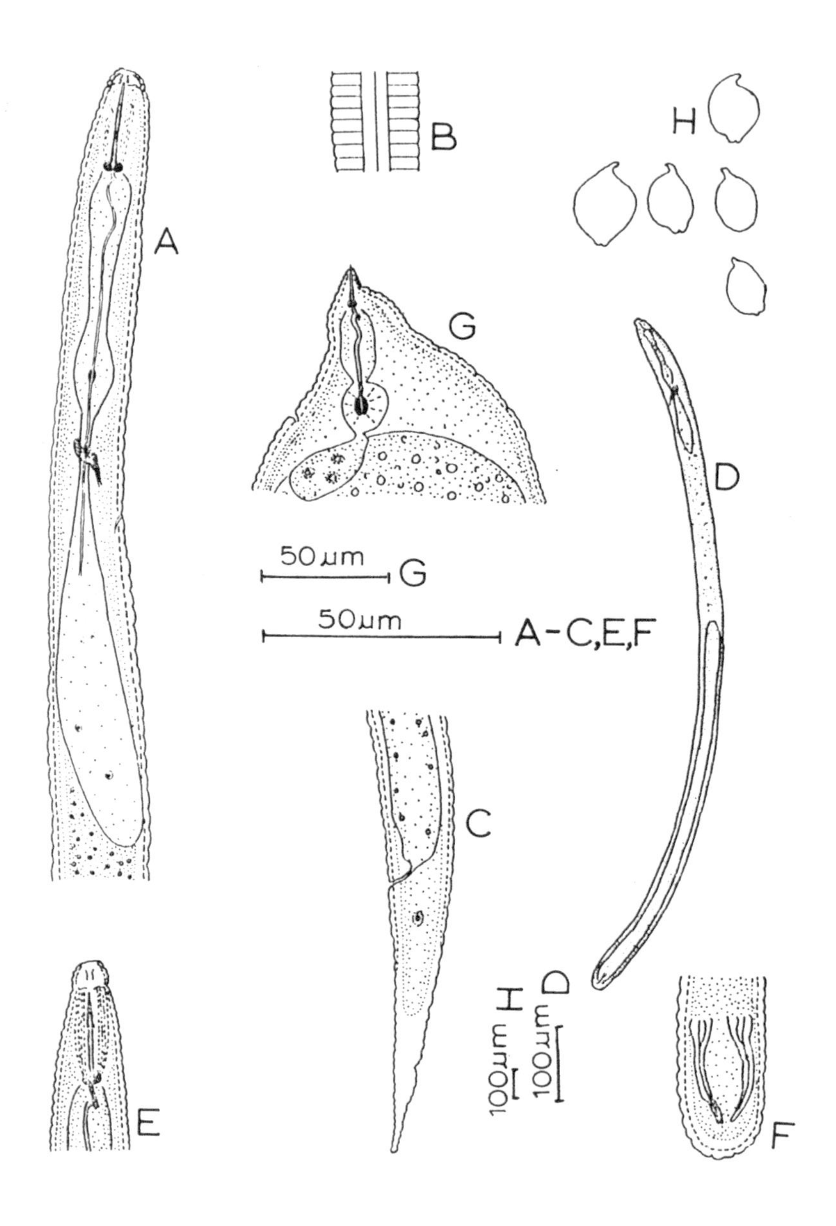

Fig. 13: *Heterodera oryzicola* Rao & Jayaprakash, 1978 (after Rao & Jayaprakash, 1978). A-D: Second stage juvenile. A – Oesophageal region; B – Lateral fields at mid-body; C – Tail; D – Entire second stage juvenile. E & F: Male: E – Head end; F – Tail. G & H: Female. G – Anterior region; H – Cyst shapes.

Cephalic sclerotization weak. Stylet strong, 18-20 μm long, basal knobs rounded. Vulva prominent, surrounded by light brown gelatinous egg sac. Anus indistinct.

Cyst: Shape same as female, light to dark brown in colour. Cuticular pattern zig-zag. Cyst cone ambifenestrate. Semifenestrae symmetrical. Fenestra 27-40 x 20-39 μm. A thin underbridge present, 84-112 x 6-20 μm, about 23 μm from fenestra. Vulval slit 36-47 μm long. The vulval supplements forming a ridge on the outer cone surface around fenestra.

Male: Body curved upon fixation. Cuticle transversely striated, 2 μm apart. Lateral fields marked by four incisures, outer bands larger. Stylet 20-30 μm long. Orifice of dorsal oesophageal gland 5 μm from stylet base. Spicules arcuate, 19-23 μm long medially. Gubernaculum 8 μm long. Cloaca with raised circular lips. Tail dorsally convex with bluntly rounded tip.

Second stage larvae: Lip region slightly offset. Cephalic framework sclerotized. Lateral fields marked with three incisures. Stylet 17-19 μm long, with rounded basal knobs. Tail 50-60 μm long, hyaline part 22-29 μm long. Phasmids on anterior third of tail.

Distinguishing characters: From *H. oryzae* it differs in the size of cyst, fenestra, female stylet, spicules, body length of second stage larvae, and a comparatively lesser hyaline part of tail.

Remarks: This species has only been reported as associated with the roots of paddy in upland fields.

21. **HETERODERA ORYZAE** LUC & BERDON, 1961
(Fig. 14)

Measurements

Females: L = 0.31-0.81 mm; breadth = 0.22-0.69 mm; length/breadth = 0.85-1.72.

Males: L = 0.79-0.97 mm; a = 33-37.6; b = 7.1-7.5; b′ = 4.9-5.5.

Second stage larvae: L = 0.37-0.50 mm; a = 22.6-28.0; b′ = 2.4-3.3; c = 5.8-6.9; c′ = 4.6-5.8; breadth = 16-20 μm.

Description

Mature Female: Body lemon-shaped, white. Cuticle striated in the anterior region. Lip region truncated with one retrorse annule and a distinct labial disc. Cephalic framework light. Stylet 28-30 μm long; basal knobs rounded posteriorly, convex anteriorly.

Cyst: Dark brown to black in colour, lemon-shaped with a well-developed vulval cone. Cuticle with zig-zag pattern; thick subcrystalline layer. Cyst cone ambifenestrate. Combined length of fenestrae 22-42 μm and width 26-43 μm. Fenestral length/width 0.60-1.05. Vulval bridge present between semifenestrae. Vulval slit 43-51 μm long. Underbridge 12-32 μm from vulval bridge, 80-120 μm long. An expanded hyaline area (40-64 x 16-35 μm) present near the middle. Usually small spherical bullae scattered inside the vulval cone. Vulval slit 42-64 μm from anus. Egg size 90-120 x 40-57 μm.

Male: Body straight or slightly ventrally curved upon fixation. Cuticle distinctly striated, 1.2-1.3 μm apart. Lateral fields marked by four incisures. Lip region dome-shaped, bearing three to four annules. Cephalic framework heavily sclerotized. Stylet strong, 23-25 μm long. Orifice of the dorsal oesophageal gland 3-5 μm from stylet base. Median oesophageal bulb ovoid. Excretory pore 119-144 μm from anterior end. Spicules curved, notched at tip, 32-36 μm long. Gubernaculum lamellate, 8-10 μm long. Phasmids not observed.

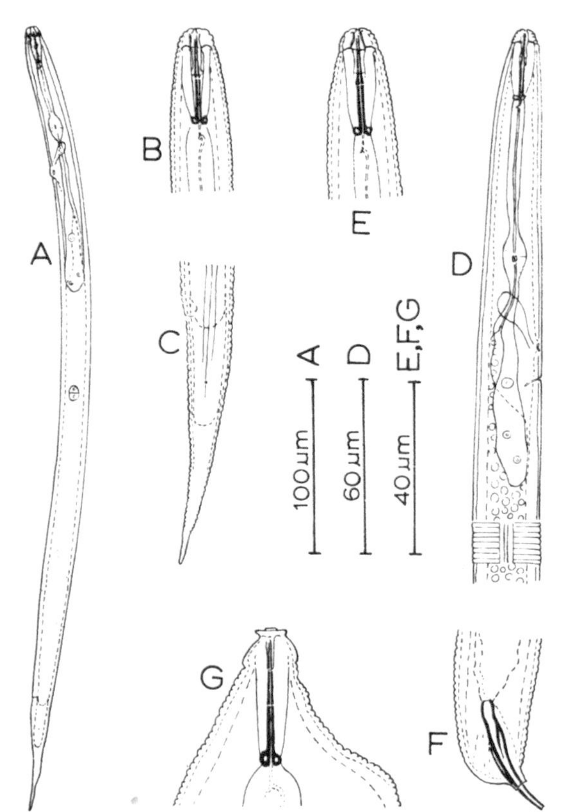

Fig. 14: *Heterodera oryzae* Luc & Berdon, 1961 (after Luc & Taylor, 1977). A-C: Second stage juvenile. A – Entire juvenile; B – Head end; C – Tail. D-F: Male. D – Anterior region; E – Head end; F – Tail. G – Anterior end of female.

Second stage larvae: Body straight or slightly ventrally curved upon fixation, gradually tapering towards both ends, more posteriorly. Cuticle transversely striated, striae 1 μm apart. Lateral fields marked by three incisures. Lip region dome-shaped with three annules. Cephalic framework heavily sclerotized. Stylet 20-22 μm long. Orifice of dorsal oesophageal gland

5-6 μm from stylet base. Median oesophageal bulb ovoid. Oesophageal glands overlapping intestine ventrally and ventrolaterally. Excretory pore 80-98 μm from anterior end. Tail conical, terminus elongate and pointed with a long hyaline terminal part. Phasmids pore-like, located in the anterior third of tail.

Distinguishing characters: Lemon-shaped cysts; vulval slit over 30 μm, presence of strong underbridge and bullae, width of fenestrae greater than length, absence of finger-like projections on the underbridge, a central expanded hyaline area; lateral fields marked by three incisures and a long tail with extensive hyaline part in second stage larvae.

Remarks: Rice is the only known host of this species. The second stage juveniles penetrate the roots of their host without any site preference. The females cause syncytia inside the vessels. The species has been recorded from Japan, Ivory Coast, India, Egypt, South America, Bangladesh, Indonesia, Iran and Philippines. Luc & Taylor (1977) have provided a detailed redescription of the species.

CYST NEMATODES
Symptoms, Loss and Control

Heterodera oryzicola has been reported as a serious pest of rice from the following states in India: Kerala, Orissa, Haryana, Madhya Pradesh and West Bengal. Kuriyan (1985) has observed 25.5-31.8% losses in the yield from Trivandrum and Quilon (Kerala).

Browning and chlorosis of leaves, browning and retardation of roots, early flowering by 10 to 13 days and partial fillings of kernel are the typical symptoms (Rao & Jayaprakash, 1977) caused by *H. oryzicola.* Cell walls may disappear near the heads of nematodes.

Jayaprakash & Rao (1983) and Rao (1985) have reported a few resistant varieties of rice against *H. oryzicola.* The soil application of phenamiphos, carbofuran and aldicarb sulfone @ 1 kg a.i./ha at seven and 50 days after transplanting are effective against *H. oryzicola.* Soaking seeds in 0.2% solution of FMC or carbofuran reduced cyst development and increased grain yield by 63-78% (Kuriyan, 1985).

Heterodera elachista Oshima, 1974 and *H. sacchari* Luc & Merny, 1963 are the other two important species causing 20-75% damage to the rice crop. The former species has been considered responsible for the failure of continuous upland rice in Japan. These species have not been reported in India.

22. CRICONEMELLA ONOENSIS (LUC, 1959) LUC & RASKI, 1981
(Fig. 15, A-C)

Syn. *Criconemoides onoense* Luc, 1959
C. onoensis Luc, 1959, cfr. Andrássy (1965)
Macroposthonia onoensis (Luc, 1959) De Grisse & Loof, 1965

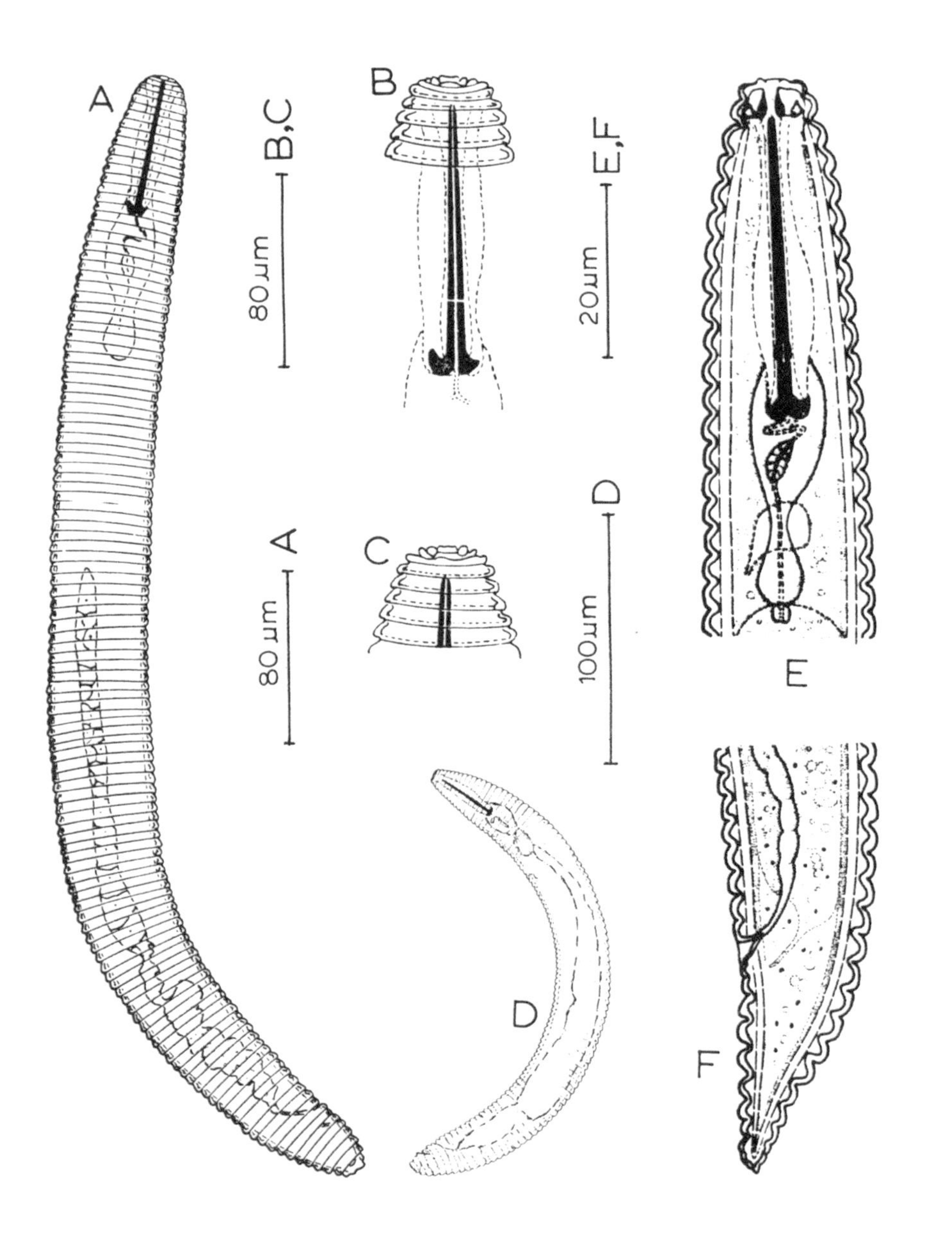

Fig. 15: *A-C: Criconemella onoensis* (Luc, 1959) Luc & Raski, 1981 (after Luc, 1970). A – Entire female; B & C – Head ends. D – Entire female of *Criconemella rustica* (Micoletzky 1915) Luc & Raski, 1981 (after de Grisse, 1969). E & F: *Hemicriconemoides cocophillus* (Loos, 1949) Chitwood & Birchfield, 1957 (after Germani & Luc, 1970). E – Anterior region; F – Posterior region.

Measurements

Females: L = 0.0389-0.670 mm; a = 10-16; b = 4.4-5.6; c = 14-20; V = 90-94; R = 111-136; Rst = 13-18; Roes = 27-31; Rex = 29-36; RV = 8-14; Ran = 7-10.

Description

Female: Body cylindrical. Cuticle marked by 111-136 annules. Lateral fields and phasmids absent. Lip region continuous with body; pseudolips greatly reduced, laterals almost absent; submedian lobes present, projecting separately. Stylet 6-10% of body length or 40-51 μm long with a very long conus and well-developed basal knobs. Female reproductive system prodelphic. Oesophagus with precorpus and a muscular postcorpus amalgamated forming enlarged cylindrical corpus, isthmus short and terminal bulb small. Vulva with open lips. Tail terminus folded on second annule, resulting in a cup-shaped end.

Remarks: Though the degree of dominance of *C. onoensis* is not high, it is abundantly found in Indian rice fields. This species has been reported as a serious pest from Guinea, Ivory Coast, West Africa and Central America, etc. About one million acres rice fields have been found infested by *C. onoensis* in Louisiana and Texas States (U.S.A). High populations in green house pots produce severe stunting, yellowing and galling of roots. Hollis & Keoboonrueng (1984) found *C. onoensis* to be widespread in rice fields causing 5-20% reduction in the yields. The application of the nematicide phenamiphos proved very effective against these nematodes.

23. **CRICONEMELLA RUSTICA** (MICOLETZKY, 1915) LUC & RASKI, 1981

(Fig. 15, D)

Syn. *Criconema rusticum* Micoletzky, 1915
Criconemoides rusticum (Micoletzky) Taylor, 1936
Criconemoides lobatus Raski, 1952
Macroposthonia rustica (Micoletzky) De Grisse & Loof, 1965

Measurements

Females: L = 0.343-0.518 mm; a = 10-14; b = 3.8-4.8; c = 19-51; V = 92-95; R = 81-107; Rst = 15-19; Roes = 25-31; Rex = 27-32; RV = 7-10; Ran = 4-9.

Description

Female: Body cylindrical. Lip region continuous. Submedian lobes present, projecting separately. Stylet 10-16% of body length or 50-60 μm long. Oesophagus typical criconematid type. Female reproductive system prodelphic. Vulva with open lips. Tail terminus mostly directed upwards and consisting of small folded annules.

Remarks: Timm & Ameen (1960) have reported this species from Bangladesh. Hollis (1967, 1969) has noted the inverse relationship between the prevalence of *Criconemella* species and rice yields.

24. **HEMICRICONEMOIDES COCOPHILLUS** (LOOS, 1949) CHITWOOD & BIRCHFIELD, 1957

(Fig. 15, E-F)

Syn. *Criconemoides cocophillus* Loos, 1914
Hemicyliophora cocophillus (Loos, 1949), Goodey, 1963
Hemicriconemoides communis Edward & Misra, 1963

Measurements

Females: L = 0.45-0.51 mm; a = 13-17; b = 4.0-5.4; c = 13-17;
V = 91-94; VL/VB = 1.2-1.7; Rst = 11-14;
R = 108-113; Ran = 8-9; Rex = 30; RVan = 1.

Description

Female: Body robust, generally straight upon fixation. Cuticular sheath attached to the inner cuticle at the anterior end but may be detached in the posterior region. Cuticle distinctly annulated, 3-5 μm wide at mid-body. Lateral fields lacking. Lip region slightly demarcated from body, marked by two to three annules (generally 2); first annule smaller than the second. Labial disc slightly elevated, may be inconspicuous in lateral view. Amphidial plate well-developed, surrounding labial disc. Stylet 50-58 μm long, with well-developed basal knobs which are 5-7 μm wide. Orifice of the dorsal oesophageal gland 5-7 μm from basal knobs. Oesophagous typical circonematid type. Female reproductive systems prodelphic. Vulval sheath well-developed. Tail convex-conoid to attenuated.

Males: Degenerate.

Remarks: *H. cocophillus* has also been found associated with rice in several districts of West Bengal (India).

25. **APHELENCHOIDES BESSEYI** CHRISTIE, 1942

(Fig. 16)

Syn. *Aphelenchoides oryzae* Yokoo, 1948

Measurements

Females: L = 0.57-0.84 mm; a = 39-53; b = 9.2-13.1; b′ = 4.06-5.77;
c = 13.8-20.4; V = 68-73.6.

Males: L = 0.53-0.61 mm; a = 40.7-46.9; b = 8.87-10.70;
b′ = 3.57-4.91; c = 16-20; T = 28-52.

Description

Female: Body straight to slightly arcuate ventrally upon fixation. Cuticle finely striated, about 1 μm apart near mid-body. Lateral fields marked by four

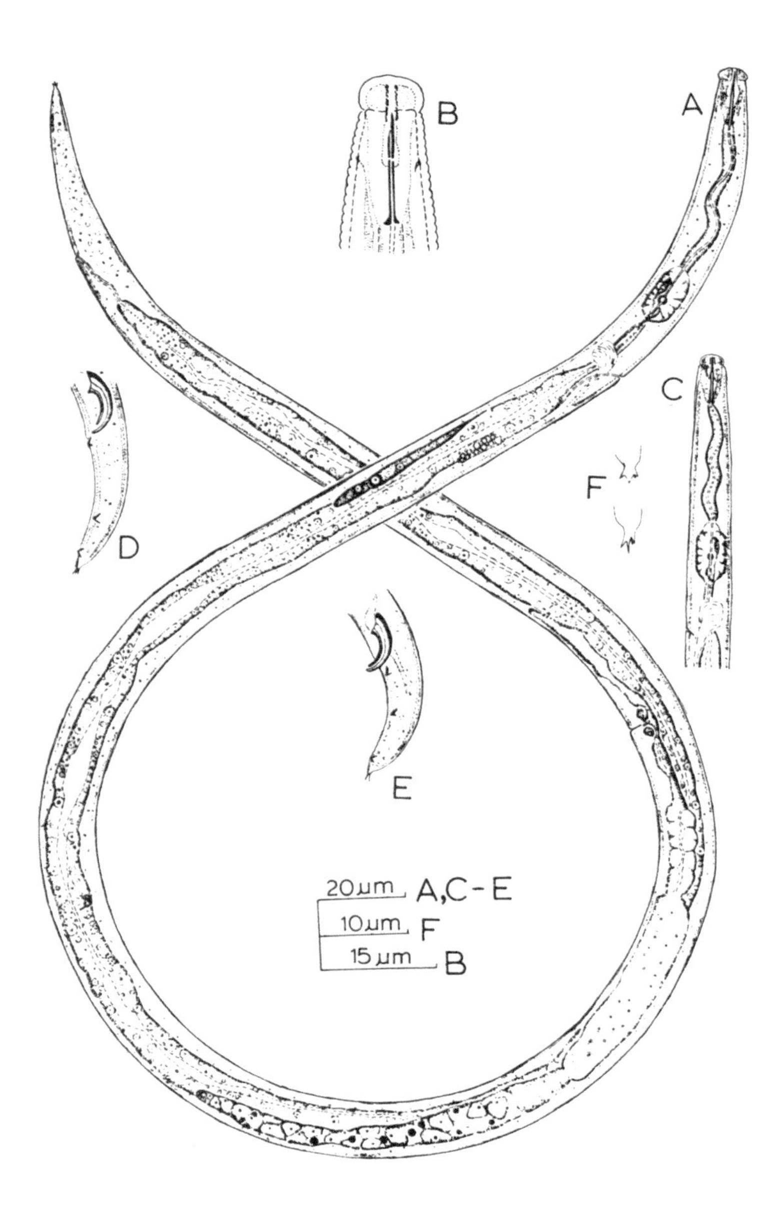

Fig. 16: *Aphelenchoides besseyi* Christie, 1942 (after Franklin & Siddiqi, 1972). A – Entire female; B – Female head end; C – Anterior region; D & E – Male tails; F – Female tail tips showing variations in shape of mucro.

incisures. Lip region rounded, unstriated, slightly wider than body. Stylet 10-13 μm long, conus about 45% of stylet length; with slight basal swellings. Median oesophageal bulb oval with a distinct valvular apparatus. Oesophageal glands overlapping intestine four to eight body widths dorsally and subdorsally. Excretory pore usually near anterior edge of nerve ring. Female reproductive system prodelphic. Ovary with oocytes in two to four rows. Spermatheca elongate oval, usually filled with sperm. Post-vulval uterine sac narrow, inconspicuous 2.5 to 3.5 times anal body width long. Tail conoid, 3.5 to five anal body-widths long, terminus with mucro of diverse shape bearing three to four pointed processes.

Male: Similar to female in general morphology except for sexual differences. Posterior end of body curved to about 180 degrees upon fixation. First pair of ventrosubmedian papillae adanal, second slightly behind middle of tail and third subterminal. Spicules typical of the genus except that the proximal end lacks a dorsal process (apex).

Remarks: Franklin & Siddiqi (1972) have provided a detailed redescription of this species.

WHITE -TIP DISEASE
Symptoms, Loss and Control

The 'white-tip disease' caused by *Aphelenchoides besseyi* was first discovered in Japan during 1940. Ou (1972) has provided a good review of this disease. The common name of the pest as 'white-tip nematode' refers to the characteristics whitening of 3-5 cm of the rice leaf tip which it causes. The white-tip nematode is reported from deep-water as well as upland rice in Africa, North America, Asia (including India and Bangladesh), Pacific and East European countries. Though rice is the natural host of *Aphelenchoides besseyi*, it can survive on other host plants as well, for example, Chinese cabbage, *Chrysanthemum,* Italian millet, onion, soybean, strawberry, sugarcane, sweet corn and sweet potato. Interestingly this nematode can also feed and complete its life cycle on many species of fungi.

Aphelenchoides besseyi is a parasite of the aerial parts of rice plants. According to Siddiqi (1980) the aphelenchs could not develop as root parasites of significance because of their weak stylet with poorly developed or without basal knobs.

The nematodes are attracted towards young seedlings or germinating seeds. First they are located inside the leaf sheath of the seedlings. As the plant grows, the nematodes move to the young growing parts of the stem and leaf on which they feed ectoparasitically. The injury caused by the stylet leads to the disintegration of phloem cells. Later, they migrate to the panicles, puncture the inflorescence and penetrate into the florets where they feed on ovary, stamens and the developing embryos. As a result of this the production of maturing grains declines or stops or sterile grains are produced.

Symptoms

Since all the symptoms caused by *A. besseyi* are in the above-ground parts of the plants, it is easy to detect the white-tip disease (see title cover). Initially the leaf tip up to 5 cm becomes light yellow to white at the tillering stage. Later these leaves become dark and finally die. The tips of flag leaves are often twisted which may check the emergence of panicles. The panicles which emerge are generally smaller than normal. The flowers become sterile and this leads to the reduction in a number of grains. Misshappen grains, stunting of the plants, later ripening and maturation, and branching from the upper nodes are commonly noted. The infested panicles are generally shorter and lighter in weight than the healthy panicles. Some of these symptoms are similar to those caused by *Ditylenchus angustus* or due to some insect injuries. It has also been observed that in the presence of white-tip nematodes yield is reduced significantly but the symptoms do not appear on the leaf and panicles.

Yield Loss

Rahman & Miah (1989) have reported that the infestation of white-tip nematode not only causes 69.5% sterile grains in panicles, the weight of grain is also reduced by 65.4%. The average loss ranges from 10-30% but in susceptible varieties the loss has been estimated up to 70%.

Life Cycle

The white-tip nematodes do not survive in soil after harvesting the crop. They generally survive up to the next crop either in the infected seeds or on weeds, ratoons etc. After harvest, they coil up and become dormant in the seeds where they may survive up to three years even if stored in dry conditions. The number of nematodes ranges from one to 64 per seed in an infested seed. When the seeds are sown in the moist soils, the nematodes are reactivated and emerge from the seed within three days. The irrigation water is the main source for spreading the nematodes.

The females lay eggs on plants. All the developmental stages (first molt-preadult) take place on the plants. The life cycle is comparatively shorter. Under optimum conditions, the life cycle may be completed within eight to 15 days. Several generations are possible within one cropping season. The optimum temperature required for the development is 23°-32°C. They become inactive if the temperature is below 13°C and humidity below 70% in the atmosphere while they may die if the temperature exceeds 43°C.

Control

Since the white-tip nematode is seed-borne, hot water treatment of the rice seeds before sowing is the best and a simple method to control them. The treatment includes soaking of seeds at 51°-56°C (without pre-soaking) for 10 to 15 minutes. The seeds should be dried properly after the treatment so that the germination is not affected, specially when the treatment is done much before sowing. For large quantities of seeds, pre-soaking for 24 hours in cold water followed by 15 minutes at 51°-56°C in water is recommended. The hot water treatment sometimes may hamper seed germination, if the temperature is not maintained precisely.

Sometimes chemical seed treatment has also been found effective against *A. besseyi,* for example, acetic thiocyanatoester REE-200, benlate 50, diazinon, malathion, parathion, phosphomidon or thiabendazole. Seeds soaked in an aqueous solution of mercuric chloride or silver nitrate and fumigation of seeds with methyl bromide have also been recommended for bringing down the number of nematodes. Successful field experiments have also been conducted by applying lebaycid, nemagon and terracur on water surface. Conventional methods like burning of straw, weeds and wild rice etc. in the field are very useful for the control of *A. besseyi.* A few rice varieties have also been tested as resistants against *A. besseyi*, for example, Century 52, Century Patna 231, Chinoor, Nira, Gumartia, and Norin No. 6, 8, 37, 39, 43.

26. **XIPHINEMA INSIGNE** LOOS, 1949

(Fig. 17, A-G)

Syn. *Xiphinema indicus* Siddiqi, 1959

Measurements

Females: L = 1.90-2.50 mm; a = 50-70; b = 5.2-8.1; c = 15-35; c′ = 3.5-8.0; V = 28-36.

Males: L = 2.10-2.30 mm; a = 52-65; b = 5.3-6.3; c = 47-54; c′ = 1.4-1.7.

Description

Female: Body long, slender, tapering gradually towards both ends. Outer cuticle smooth, inner cuticle finely striated transversely, thickest at the vulva and tail regions. Lateral chords about one-fourth of body-width near the middle. Lip region slightly offset, almost flat to rounded anteriorly. Amphids stirrup shaped; their aperture occupying about three-fifths of the corresponding body width. Odontostyle 80-111 μm or about 7-10 labial widths long. Odontophore 55-64 μm long. Fixed guiding ring 7-10 labial widths from anterior extremity. Basal expanded part of oesophagus occupying about 20% of neck region. Cardia short and conoid. Prerectum 18 to 20 times the anal body width long. Rectum 1.2 to 1.5 anal body-widths long. Female reproductive system amphidelphic, comprising the usual parts, uterus convoluted. The anterior ovary generally reduced. Vagina about half of the corresponding body width long. Tail narrow, conoid to elongate filiform, 3.5 to eight times of anal body width, with three to four caudal pores on each side.

Male: Similar to females in general shape and morphology except for the reproductive system, more curved posterior region and shorter tail. Odontostyle 93-104 μm long. Spicules sharply ventrally curved. Lateral guiding pieces 15-16 μm long. Supplements consisting of an adanal pair and three to five irregularly spaced ventromedians.

Remarks: This is a widely distributed and highly variable species. Bajaj & Jairajpuri (1977) have provided intraspecific variations of *X. insigne*. This and some other *Xiphinema* spp. are found in paddy fields occasionally, but their role as pathogens is not fully known, though they cause severe damage to other crops and may also act as vectors of soil-borne diseases.

27. PARATRICHODORUS POROSUS (ALLEN, 1957) SIDDIQI, 1974 (Fig. 17, H-L)

Syn. *Trichodous porosus* Allen, 1957

Measurements

Females: L = 0.46-0.77 mm; a = 15-25; b = 4.1-5.5; c = subterminal; $V = {}^{19\text{-}28}53\text{-}58^{12\text{-}25}$

Males: L = 0.53-0.77 mm; a = 15-25; b = 4.0-6.0; c = 60- 90; T = 55-74.

Description

Female: Body robust, tapering gradually towards the anterior end. Cuticle smooth, sometimes appearing to be loose or detached from body to fixation. Lip region rounded. Onchiostyle typical of the genus, 43-50 μm long. Excretory pore at the level of mid-oesophagus. The anterior slender part of oesophagus expanding gradually to form terminal bulb. Intestine overlapping posterior end of the terminal bulb. Female reproductive system amphidelphic, comprising the usual parts. A pair of ventromedian pores present on either side (anterior and posterior) of the vulva. Vagina slightly sclerotized distally.

Male: Similar to females in general shape and morphology except for the reproductive system. Spicules straight, 36-39 μm long. Gubernaculum 12-13 μm long with a slightly thickened distal end. Two ventromedian supplements present in the spicular region. Two postanal ventrosubmedian papillae also present. Tail short, rounded, with terminal caudal pores.

Remarks: This species has been found from the soil around roots of rice in West Bengal.

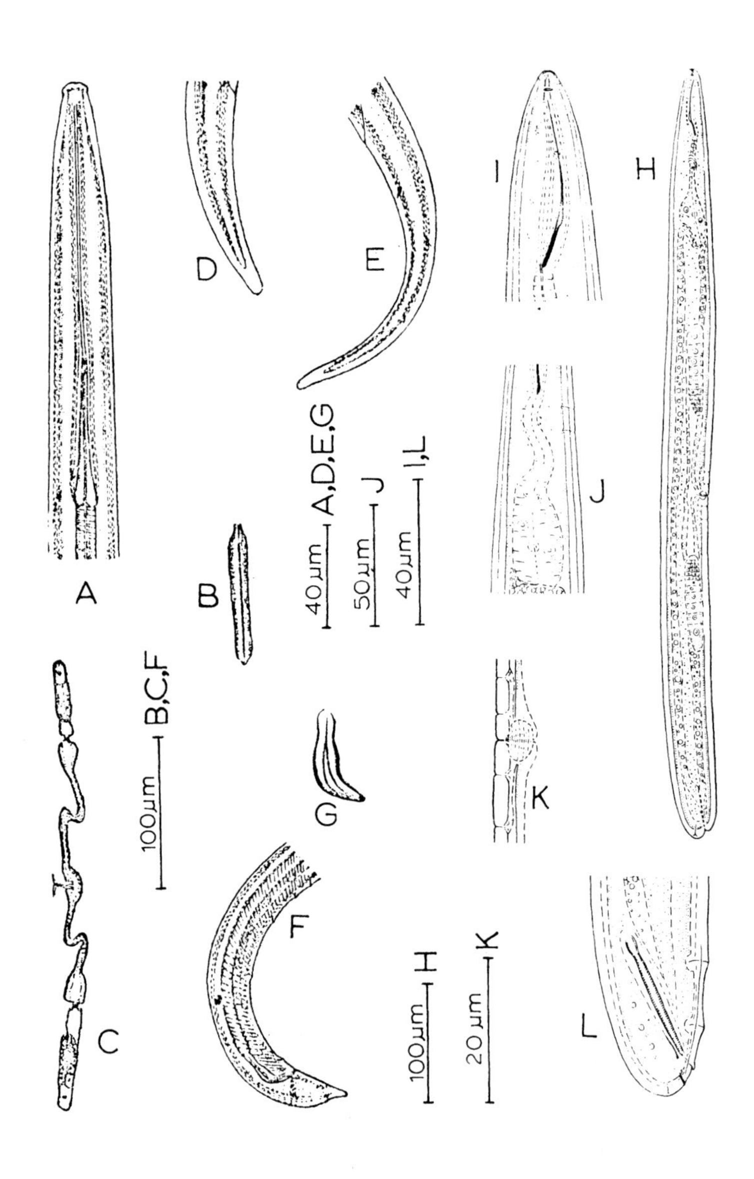

Fig. 17: A-G: *Xiphinema insigne* Loos, 1949 (after Bajaj & Jairajpuri, 1977). A – Anterior region; B – Basal expanded part of oesophagus; C – Female reproductive system; D & E – Female tails; F – Male posterior region; G – Spicule. H-L: *Paratrichodorus porosus* (Allen, 1957) Siddiqi, 1974 (after Allen, 1957). H – Entire female; I – Female head end; J – Oesophageal region of male; K – Vulva region showing ventromedian pores; L – Male tail.

IDENTIFICATION KEY TO NEMATODES OF PADDY

1. Both sexes vermiform 5

   Females swollen, saccate, males vermiform 11

2. Ectoparasites of the above-ground parts of paddy 3

   Ecto- or endoparasites of roots of paddy 4

3. Nematodes causing 'white-tip' disease; oesophagus with enlarged postcorpus (median bulb) occupying three-fourths of body width, orifice of dorsal oesophageal gland in the postcorpus just above the valvular apparatus .... *Aphelenchoides besseyi*
   Nematodes causing 'ufra' disease; oesophagus with normal-sized postcorpus (median bulb), orifice of dorsal oesophageal gland in the procorpus a little behind the stylet ....
   .... *Ditylenchus angustus*

4. Oesophagus bipartite, consisting of a slender part and a basal expanded portion 5
   Oesophagus tripartite, consisting of procorpus, median bulb and a basal bulb/lobe 6

5. Body slender, more than 1.5 mm long; cuticle finely striated; feeding apparatus odontostyle, long needle-like with thin lumen, followed by odontophore; female tail rounded to elongate-conoid
   .... *Xiphinema* spp.
   Body robust, less than 1 mm long, cuticle smooth; feeding apparatus onchiostyle, curved, anteriorly solid and posteriorly hollow, female tail very short to absent ....
   .... *Paratrichodorus* spp.

6. Body ring-like with crenate annules, pro- and postcorpus amalgamated forming an enlarged cylindroid corpus .... (i)

   (i) Female body enclosed in a cuticular sheath ....
   .... *Hemicriconemoides cocophillus*
   Female body without cuticular sheath .... (ii)

   (ii) Female stylet length 40-51 μm; R = 111-136; RV = 8-14
   .... *Criconemella onoensis*
   Female stylet length 50-60 μm; R = 81-107; RV = 7-10
   .... *Criconemella rustica*

   Cuticle transversely striated, body without ring-like or crenate annules; pro- and postcorpus distinctly demarcated 7

7. Oesophageal glands enclosed in a terminal bulb, offset from intestine; oesophago-intestinal junction with well developed cardia .... *Tylenchorhynchus* spp.

   Males common; spermatheca functional .... *T. mashhoodi*

   Males absent; spermatheca non-functional .... *T. annulatus*

   Oesophageal glands not enclosed in a bulb but overlapping intestine, separate or fused; oesophago-intestinal junction marked by a valve 8

8. Phasmids scutella-like, one anterior and the other posterior to vulva (not opposite each other) .... *Hoplolaimus indicus*

   Phasmids pore-like, on or a little pre-anal (opposite each other) 9

9. Body spiral upon death; lip region high; orifice of dorsal oesophageal gland situated considerably posterior to base of stylet .... *Helicotylenchus* spp.

   (i) Female tail with a non-annulated ventral projection or a narrow terminus .... (ii)

   Female tail with an annulated ventral projection and a distinct indentation on dorsal side enveloped by cuticular fold .... *H. crenacauda*

   (ii) Stylet length 24-27 μm .... *H. dihystera*

   Stylet length 21-22 μm .... *H. abunaamai*

   Body almost straight or slightly ventrally curved upon death; lip region low; orifice of dorsal oesophageal gland near base of stylet 10

10. Body more than 1.4 mm long; tail elongate conoid; female reproductive system amphidelphic .... *Hirschmanniella* spp.

    (i) Stylet length 40-50 μm .... *H. spinicaudata*

    Stylet length less than 32 μm .... (ii)

    (ii) Stylet length 29-32 μm .... *H. imamuri*

    Stylet length less than 29 μm .... (iii)

    (iii) Stylet length 24-29 μm .... *H. mucronata*

    Stylet length less than 24 μm .... (iv)

(iv) Tail in both sexes with a ventral notch near terminus .... *H. caudacrena*

Tail without ventral notch near terminus .... (v)

(v) Spermatheca inconspicuous, without sperm; males rare .... *H. belli*

Spermatheca conspicuous, with sperm; males abundant .... (vi)

(vi) Stylet length 16-19 μm .... *H. oryzae*

Stylet length 20-24 μm .... *H. gracilis*

Body length less than 0.8 mm; tail cylindrical to subcylindrical; female reproductive system mono-prodelphic .... *Pratylenchus* spp.

Lip region marked by two annules; spermatheca filled with sperm .... *P. coffeae*

Lip region marked by three annules; spermatheca without sperm .... *P. thornei*

11. Females with only neck inside roots, transforming into cyst upon maturity .... *Heterodera* spp.

Female stylet 28-30 μm long; fenestra length 40-50 μm .... *H. oryzae*

Female stylet 18-20 μm long; fenestra length 27-40 μm .... *H. oryzicola*

Females within roots, forming knots .... *Meloidogyne* spp.

(i) Female stylet length 10-11 μm .... *M. graminicola*

Female stylet length more then 14 μm .... (ii)

(ii) Lip annules two to three behind head cap; conus of stylet distinctly dorsally curved and shaft slightly wider near basal knobs .... *M. incognita*

Lip annule one behind head cap; conus of stylet slightly curved dorsally and shaft cylindrical .... *M. javanica*

OTHER PLANT PARASITIC NEMATODES FROM RICE FIELDS

ORDER TYLENCHIDA THORNE, 1949

FAMILY TYLENCHIDAE ÖRLEY, 1880

Tylenchus hayati Khan, 1985
Filenchus filiformis (Bütschli, 1873) Meyl, 1961
Coslenchus costatus (de Man, 1921) Siddiqi, 1978
Basiria elegans (Khan & Khan, 1975) Bajaj & Bhatti, 1979
B. tumida (Colbran, 1960) Geraert, 1968
Neopsilenchus magnidens (Thorne, 1949) Thorne & Malek, 1968

FAMILY ANGUINIDAE NICOLL, 1935

Ditylenchus mirus Siddiqi, 1963
Nothotylenchus acutus Husain & Khan, 1968

FAMILY DOLICHODORIDAE CHITWOOD IN CHITWOOD & CHITWOOD, 1950

Brachydorus tenuis De Guiran & Germani, 1968

FAMILY TYLENCHORHYNCHIDAE ELIAVA, 1964

Tylenchorhynchus claytoni Steiner, 1937
Divittus divittatus (Siddiqi, 1961) Jairajpuri, 1984
Mulkorhynchus phaseoli (Sethi & Swarup, 1968) Jairajpuri, 1988
Ulignotylenchus palustris (Merny & Germani, 1968) Siddiqi, 1971
U. rhopalocercus (Seinhorst, 1963) Siddiqi, 1971
Paratrophurus costarriensis Lopez, 1968
Trichotylenchus falciformis Whitehead, 1960

FAMILY HOPLOLAIMIDAE FILIPJEV, 1934

Hoplolaimus columbus Sher, 1963
Helicotylenchus exallus Sher, 1966
H. erythrinae (Zimmermann, 1904) Golden, 1956
H. indicus Siddiqi, 1963
H. pseudorobustus (Steiner, 1914) Golden, 1956
H. retusus Siddiqi & Brown, 1964
Rotylenchus gracilidens (Sauer, 1958) Sauer, 1958
R. unisexus Sher, 1965
Scutellonema brachyurum (Steiner, 1938) Andrássy, 1958

FAMILY ROTYLENCHULIDAE HUSAIN & KHAN, 1976

Rotylenchulus reniformis Linford & Oliveira, 1940

FAMILY PRATYLENCHIDAE THORNE, 1949

Pratylenchus brachyurus (Godfrey, 1929) Filipjev & Sch. Shek., 1941
P. indicus Das, 1960

P. scribneri Steiner, 1943
P. zeae Graham, 1951
Hirschmanniella shamimi Ahmad, 1974
H. thornei Sher, 1968

FAMILY MELOIDOGYNIDAE SKARBILOVICH, 1959
Meloidogyne acrita Chitwood, 1941
M. arenaria (Neal, 1889) Chitwood, 1949
M. exigua Goeldi, 1982
M. oryzae Mass, Sanders & Dede, 1978
M. salasi Lopez, 1984

FAMILY HETERODERIDAE FILIPJEV & SCH. STEK., 1941
Heterodera elachista Oshima, 1974
H. graminophila Golden & Birchfield, 1972
H. saccharri Luc & Merny, 1963

FAMILY CRICONEMATIDAE TAYLOR, 1936
Criconema corbetti (De Grisse, 1967) Raski & Luc, 1985
Criconemella crenata (Loof, 1964) Luc & Raski, 1981
C. curvata (Raski, 1958) Luc & Raski, 1981
C. incisa (Raski & Golden, 1966) Luc & Raski, 1981
C. obtusicaudata (Heyns, 1962) Luc & Raski, 1981
C. onosteris (Phukan & Sanwal, 1981)
C. ornata (Raski, 1958) Luc & Raski, 1981
C. reedi (Diab & Jenkins, 1966) Luc & Raski, 1981
C. sphaerocephala (Taylor, 1936) Luc & Raski, 1981
Hemicriconemoides brachyurus (Loos, 1949) Chitwood & Birchfield, 1957

FAMILY HEMICYCLIOPHORIDAE SKARBILOVICH, 1959
Hemicycliophora oryzae De Waele & van den Berg, 1988
H. typica de Man, 1921
Caloosia heterocephala Rao & Mohandas, 1976

FAMILY PARATYLENCHIDAE THORNE, 1949
Paratylenchus aquaticus Merny, 1966
P. dianthus Jenkins & Taylor, 1956
Gracilacus janai Baqri, 1979

ORDER APHELENCHIDA SIDDIQI, 1980
FAMILY APHELENCHIDAE FUCHS, 1937
Aphelenchus avenae Bastian, 1865

FAMILY APHELENCHOIDIDAE SKARBILOVICH, 1947
Aphelenchoides asterocaudatus Das, 1960

A. bicaudatus (Imamura, 1931) Filipjev & Sch. Stek., 1941
A. saprophilus Franklin, 1957
A. subtenuis (Cobb, 1926) Steiner & Buhrer, 1932

ORDER DORYLAIMIDA PEARSE, 1942
FAMILY LONGIDORIDAE THORNE, 1935
Longidorus pisi Edward, Misra & Singh, 1964
Paralongidorus australis Sterling & McCulloch, 1984
P. citri (Siddiqi, 1959) Siddiqi, Hooper & Khan, 1963

FAMILY XIPHINEMATIDAE DALMASSO, 1969
Xiphinema brevicolle Lordello & Costa, 1961
X. orbum Siddiqi, 1964
X. oryzae Bos & Loof, 1985

FAMILY NORDIIDAE JAIRAJPURI & A.H. SIDDIQI, 1964
Longidorella parva Thorne, 1939
Lenonchium macrodorus Ahmad & Jairajpuri, 1988
L. oryzae Siddiqi, 1965

ORDER TRIPLONCHIDA COBB, 1920
FAMILY TRICHODORIDAE THORNE, 1935
Trichodorus petrusalberti De Waele, 1988
Paratrichodorus lobatus (Colbran, 1965) Siddiqi, 1974

NEMATODES FOUND FROM SOIL AROUND ROOTS OF RICE

ORDER DORYLAIMIDA PEARSE, 1942

Dorylaimus stagnalis Dujardin, 1845
D. innovatus Jana & Baqri, 1983
D. siddiqii Ahmad & Jairajpuri, 1982
D. thornei Andrássy, 1969
Ischiodorylaimus novus Baqri & Jana, 1986
Mesodorylaimus globiceps Loof, 1964
M. kowyni Basson & Heyns, 1974
Drepanodorylaimus flexus (Thorne & Swanger, 1936) Andrássy, 1969
Laimydorus agilis (de Man, 1880) Siddiqi, 1969
L. baldus Baqri & Jana, 1982
L. dhanachandi Jairajpuri & Ahmad, 1983
L. distinctus Dey & Baqri, 1986
L. oryzae Dey & Baqri, 1986
L. siddiqii Baqri & Jana, 1982
Calodorylaimus andrassyi Baqri & Jana, 1982
C. simplex Baqri & Jana, 1982
Thornenema coomansi (Baqri & Jana, 1980) Carbonell & Coomans, 1987
T. mauritianum (Williams, 1959) Baqri & Jairajpuri, 1967
T. oryzae (Ahmad & Jairajpuri, 1982) Carbonell & Coomans, 1987
T. pseudosartum Carbonell & Coomans, 1987
T. qaiseri Sauer, 1981
T. shamimi (Baqri & Jana, 1980) Carbonell & Coomans, 1987
Opisthodorylaimus cavalcantii (Lordello, 1955) Carbonell & Coomans, 1986
Ecumenicus monhystera (de Man, 1880) Thorne, 1974
Labronemella andrassyi (Baqri & Khera, 1976) Andrássy, 1985
Crateronema aestivum Siddiqi, 1969
Discolaimoides bulbiferous (Cobb, 1906) Heyns, 1963
D. filiformis Das, Khan & Loof, 1969
Aporcelaimellus chauhani Baqri & Khera, 1975
A. heynsi Baqri & Jairajpuri, 1968
A, tropicus Jana & Baqri, 1981
Belondira nepalensis Siddiqi, 1964
Dorylaimellus indicus Siddiqi, 1964
D. deviatus Baqri & Jairajpuri, 1968
D. discocephalus Siddiqi, 1964
Paraoxydirus gigas (Jairajpuri, 1964) Jairajpuri & Ahmad, 1979

Tylencholaimus obscurus Jairajpuri, 1965
T. pakistanensis Timm, 1964
T. paradoxus Loof & Jairajpuri, 1968
Discomyctus cephalatus Thorne, 1939
Tantunema aquaticum Ahmad & Jairajpuri, 1988
Proleptonchus clarus Timm, 1964
P. indicus Siddiqi & Khan, 1964
Dorylaimoides arcuatus Siddiqi, 1964
D. arcuicaudatus Baqri & Jairajpuri, 1969
D. constrictus Baqri & Jairajpuri, 1969
D. elaboratus Siddiqi, 1965
D. kalingus Ahmad & Jairajpuri, 1983
D. leptura Siddiqi, 1965
D. parvus Thorne & Swanger, 1936
D. teres Thorne & Swanger, 1936
Morasia bengalensis Jana & Baqri, 1981
Tyleptus variabilis Jairajpuri & Loof, 1964
Basirotyleptus minimus Jana & Baqri, 1981
Neoactinolaimus elaboratus (Cobb, 1906) Heyns & Agro, 1969
N. thornei Chaturvedi & Khera, 1979
Laevides imphalus Ahmad & Jairajpuri, 1980

ORDER MONONCHIDA

Mononchus aquaticus Coetzee, 1968
Mylonchulus lacustris (N.A. Cobb in M.V. Cobb, 1915) Andrássy, 1958
M. minor (Cobb, 1893) Andrássy, 1958
Paramylonchulus mulveyi (Jairajpuri, 1970) Jairajpuri & Khan, 1982
Miconchus aquaticus Khan, Ahmad & Jairajpuri, 1978
Iotonchus trichurus (Cobb, 1916) Andrássy, 1958
Mononchulus nodicaudatus (Daday, 1901) Schneider, 1937
Oionchus obtusus Cobb, 1913
O. paraobtusus Jairajpuri & Khan, 1982

REFERENCES

AHMAD, N., P.K. DAS, & Q.H. BAQRI. Evaluation of yield losses in rice due to *Hirschmanniella gracilis* (de Man, 1888) Luc & Goodey, 1963 (Tylenchida: Nematoda) at Hooghly (West Bengal). *Bull. zool. Surv. India. 5*: 85-91, 1984.

AHMAD, W. & M.S. JAIRAJPURI. Studies on the genus *Lenonchium* (Nematoda: Dorylaimida) with description of *L. macrodorus* n.sp. *Revue Nematol. 11*: 7-11, 1988.

AHMAD, W., T.H. KHAN, & A.L. BILGRAMI. Plant parasitic nematodes associated with paddy crop in Bihar, India. *Int. Nematol. Network Newsl. 5*: 4, 1988.

AMEEN, M. Plant parasitic nematodes of the subfamily Longidorinae in East Pakistan. *Proc. Pak. Sci. Cong. 12th.* Part III, Sec. B, 25, 1960.

BAJAJ, H.K. & M.S. JAIRAJPURI. Variability within *Xiphinema insigne* populations from India. *Nematologica. 23*: 33-46, 1977.

BAKR, M.A. Occurrence of ufra disease in transplanted rice. *Int. Rice Res. Newslt. 3*: 16, 1977.

BANERJEE, S.N. & D.K. BANERJEE. Occurrence of the nematode *Hoplolaimus indicus* in West Bengal. *Curr. Sci. 35* (23): 597-598, 1966.

BAQRI, Q.H. & N. AHMAD. Nematodes from West Bengal (India) XVI. On the species of the genus *Helicotylenchus* Steiner, 1945 (Hoplolaimidae: Tylenchida). *J. Zool. Soc. India. 35*: 24-48, 1984

BAQRI, Q.H., N. AHMAD & S. DEY. Nematodes from West Bengal (India). XXIV. Qualitative and quantitative studies of plant and soil inhabiting nematodes associated with paddy crop in Coochbehar. *Rec. Zool. Surv. India. 88*: 63-69, 1991.

BAQRI, Q.H. & P.K. DAS. Nematodes from West Bengal (India). XIX. Qualitative and Quantitative studies of plant and soil inhabiting nematodes associated with paddy crop in West Dinajpur district. *Rec. Zool. Surv. India*. (in press), 1991.

BAQRI, Q.H., S. DEY & S. GHOSH. Effect of different sources of nitrogen on the management of *Hirschmanniella gracilis* (de Man, 1880) Luc & Goodey, 1963 (Tylenchida: Nematoda) associated with paddy crop. *Indian J. Helminth.* (n.s.), *4*: 12-20, 1988.

BAQRI, Q.H. & S. DEY. Nematodes from West Bengal (India). XXIII. Qualitative and quantitative studies of plant and soil inhabiting nematodes associated with paddy crop in district Darjeeling. *Rec. Zool. Surv. India.* (in press), 1991.

BAQRI, Q.H. & M.S. JAIRAJPURI. On the intraspecific variations of *Tylenchorhynchus mashhood* Siddiqi & Basir, 1959 and emended key to species of *Tylenchorhynchus* Cobb, 1913. *Rev. Brasil. Biol. 1*: 61-68, 1970.

BAQRI, Q.H., A. JANA, N. AHMAD & P.K. DAS. Nematodes from West Bengal (India) VIII. Qualitative and quantitative studies of plant and soil inhabiting nematodes associated with paddy crop in Burdwan district. *Rec. Zool. Surv. India. 80*: 331-340, 1983.

BARAT, H., M. DELASSUS & H.H. VOUNG. The geographical distribution of white tip disease of rice in tropical Africa and Madagascar. In "Nematodes of tropical crops" (Ed. J.E. Peachey). *Tech. Commun. Bur. Helminth. 40*: 269-273, 1969.

BIRAT, R.B.S. New records of parasitic nematodes of rice (*Oryza sativa* L.) in Bihar. *Sci. Cult. 31*(9): 494, 1965.

BISWAS, H. & RAO, Y.S. Studies on nematodes of rice and rice soils – II. Influence of *Meloidogyne graminicola* on yield of rice. *Oryza, 8*: 101-102, 1971.

BUTLER, E.J. Ufra disease of rice. *Agric. J. India. 8*: 205-220, 1913a.

BUTLER, E.J. Disease of rice. An eelworm disease of rice. *Agr. Res. Inst. Pusa. Bull. 34* B: 1-27, 1913b.

BUTLER, E.J. The rice worm (*Tylenchus angustus*) and its control. *Botanical Series. X*: 1-37, 1919.

CATLING, H.D., S. ALAM & S.A. MIAH. Assessing losses in rice due to insects and disease in Bangladesh. *Expt. Agric. 14*: 1-11, 1978.

CATLING, H.D., P.G. COX, Z. ISLAM & L. RAHMAN. Two destructive pests of deep-water rice yellow stem borer and ufra. *THE ADAB News. VI*(8): 16-21, 1979.

CHAWLA, M.L. Ph.D. thesis submitted to P.G. School, I.A.R.I., New Delhi, 1972.

COX, P.G. Symptoms of ufra disease of deep-water rice in Bangladesh. *Int. Rice Res. Newsl. 5*: 18, 1980.

COX, P.G. & L. RAHMAN. The ufra nematode population in deep-water rice in Bangladesh. *Int. Rice Res. Newsl., 4*(3): 10-11, 1979a.

COX, P.G. & L. RAHMAN. Synergy between benomy and carbofuran in the control of ufra. *Int. Rice Res. Newsl., 4*(4): 11, 1979b.

COX, P.G. & L. RAHMAN. Effect of ufra disease on yield loss of deep-water rice in Bangladesh. *Tropical Pest Management. 26*: 410-415, 1980.

DAS, P.K., N. AHMAD & Q.H. BAQRI. The study on seasonal variations in the population of *Hirschmanniella gracilis* (de Man, 1880) Luc & Goodey, 1963 (Tylenchida: Nematoda). *Indian J. Helminth.* (n.s.), 1: 17-25, 1984.

DEY, S. & Q.H. BAQRI. Nematodes from West Bengal (India). XX. Morphometric and allometric variations in *Hirschmanniella gracilis* (de Man, 1880) Luc & Goodey, 1963 (Radopholidae: Tylenchida). *Indian J. Helminth* (n.s.), *2*: 71-80, 1985.

FORTUNER, R. Nematode root parasites associated with rice in Senegal (High Cosamance and Central and Northern regions) and in Mauritania. *Ser. Biol. Nematol. 10*: 147-159, 1975.

FORTUNER, R. Root parasitic nematodes associated with rice in Senegal. *Thesis Univ. Clande Bernard Lyon, France. 50,* 1976.

FORTUNER, R. & G. MERMY. Root-parasitic nematodes of rice. *Revue Nematol. 2*: 79-102, 1979.

FRANKLIN, M.T. & M.R. SIDDIQI. *Aphelenchoides besseyi*. CIH Descriptions of Plant-parasitic Nematodes. Set 1, No. 4. *Farnham Royal, U.K. Commonwealth Agricultural Bureaux,* 1972.

GOLDEN, A.M. & W. BIRCHFIELD. Rice root knot nematode *Meloidogyne graminicola* as a new pest of rice. *Pl. Dis. Reptr. 52*: 423, 1968.

GONZALEZ, F.C. Plant parasitic nematodes associated with rice and corn in several agricultural areas of Costa Rica. *Agron. Costa. 2*: 171-173, 1978.

HASHIOKA, Y. The rice stem nematode *Ditylenchus angustus* in Thailand. *Plant Prot. Bull. FAO, 11*(5): 97-102, 1963.

HASHIOKA, Y. Nematode disease of rice in the world. *Riso, 13*: 139-147, 1964.

HOLLIS, J.P. Concentration of toxicants in the rice soil profile (abstract). *Phytopathology. 57*: 460, 1967.

HOLLIS, J.P. Genesis of soil nematode problem in Louisiana rice. *Int. Rice Comm. Newsletter. 18*: 19-28, 1969.

HOLLIS, J.P. The ring nematode problem in rice. *La Agric. 23*: 6-7, 1980.

HOLLIS, J.P. & S. KEOBOONRUENG. Nematode parasites of rice. *Plant and Insect Nematodes* (Ed. W.R. Nickle). Marcel Dekker, Inc., N.Y., Basel.: 95-146, 1984.

ICHINOHE, M. 6. Nematode disease of rice. *Economic Nematology* (Ed. J.M. Webester). Academic Press, London, N.Y.: 127-143, 1972.

JACQ, V.A. & R. FORTUNER. Biological control of rice nematodes using sulphate reducing bacteria. *Revenue Nematol. 2*: 41-50, 1979.

JAIRAJPURI, M.S. & W.U. KHAN. *Predatory nematodes (Mononchida)*. Associated Publishing Company, New Delhi, India. 131 pp. 1982.

JAYAPRAKASH, A. & Y.S. RAO. Reaction of rice cultivators to the cyst nematode, *Heterodera oryzicola. Indian J. Nematol. 13*: 117-118, 1983.

JENA, R.N. & Y.S. RAO. Root knot nematode resistance in rice. *Indian J. Genetic. Plt. Brdg. A. 34*: 443-449, 1973.

JOSHI, M.M. & J.P. HOLLIS. Pathogenecity of *Tylenchorhynchus martini* swarmers to rice. *Nematologica. 22*: 123-124, 1976.

KHAN, E. & M.L. CHAWLA. *Hoplolaimus indicus*. CIH Descriptions of Plant-Parasitic Nematodes. Set 5, No. 66. *Farnham Royal, U.K. Commonwealth Agricultural Bureaux*, 1975.

KHUONG, H.B. *Hirschmanniella* spp. in rice fields of Vietnam. *J. Nematol. 19*: 82-84, 1987.

KURIYAN, K.J. Biennial Report of the All India Coordinated Research Project on Nematode Pests of Agricultural Crops and Their Control (ICAR), Kerala, 1983-85.

KURIYAN, K.J., U. KUMARI & R. HEBSYBAI. Rice cyst nematode, *Heterodera oryzicola* in soils of Kerala State (Abst.). *Proc. 4th Nematol. Symp., Nematol. Soc. India.* Udaipur, 1985.

LIN, Y.Y. Studies on the rice nematodes in Taiwan. *J. Agric. Forest Chung-Hsing Univ. 19*: 13-27, 1970.

LUC, M. & R. FORTUNER. *Hirschmanniella spinicaudata*. CIH Descriptions of Plant-Parasitic Nematodes. Set 5, No. 68. *Farnham Royal U.K. Commonwealth Agricultural Bureaux*, 1975.

LUC, M. & D.P. TAYLOR. *Heterodera oryzae.* CIH Descriptions of Plant-Parasitic Nematodes. Set 7, No. 91. *Farnham Royal, U.K. Commonwealth Agricultural Bureaux*, 1977.

MALIK, R. & Z. YASMIN. Nematodes in paddy fields of Pakistan. *Int. Rice Res. Newsl. 3*: 16-17, 1978.

MATHUR, V.K. & S.K. PRASAD. Occurrence and distribution of *Hirschmanniella oryzae* in the Indian Union with description of *H. mangalorensis* sp.n. *Indian J. Nematol. 1*: 220-226, 1971.

MATHUR, V.K. & S.K. PRASAD. Control of *Hirschmanniella oryzae* associated with paddy. *Indian J. Nematol.* (1973), 3: 54-60, 1974.

MERNY, G. Les nématodes phytoparasites des riziérés inondées en cote d'ivoire I.-les espices observes. *Cash. ORSTOM, Sir. Biol. n. 11*: 3-45, 1970.

MERNY, G. The phytoparasitic nematodes of the flooded rice paddies of the Ivory Coast. III. Studies on the population dynamics of two endoparasites. *Hirschmanniella spinicaudata & H. oryzae. Cash. ORSTOM. Ser. Biol. 16*: 31-87, 1972.

MIAH, S.A. Disease problems and progress of research on ufra disease of rice in Bangladesh. *IRC NEWSL. 33*(2): 35-38, 1984.

MIAH, S.A. & M.A. BAKR. Nematode disease of root. In Lit. Rev. of Insect pests and diseases of rice. BRRI, 125-129, 1977.

MIAH, S.A. & M.L. RAHMAN. Observation on incidence and chemical control of ufra in Boro fields. *Int. Rice. Res. Newsl. 6:* 12, 1981.

MIAH, S.A., A.K.M. SAHJAHAN, N.R. SHARMA, A.L. SHAH, & N.I. BHUIYAN. Interaction of zinc deficiency and ufra disease on rice in Bangladesh. *Bangladesh J. Agric. 9:* 61-67, 1984.

McGEACHIE, I. & M.L. RAHMAN. Ufra disease: A review and a new approach to control. *Tropical Pest Management. 29*: 325-332, 1983.

MONDAL, A.H., M.L. RAHMAN & S.A. MIAH. Interaction between ufra sheath rot disease of rice (Abst.). *Proc. 10th Ann. Bangladesh Sci. Conf. Dhaka.* March 22-27, 1985.

MONDAL, A.H., A. RAHMAN, H.U. AHMAD & S.A. MIAH. The causes of increasing blast susceptibility of ufra infested rice plants. *Bangladesh J. Agril. 11*: 77-79, 1986.

MONDAL, A.H., M.L. RAHMAN & S.A. MIAH. Yield loss due to transplanting of ufra infested seedlings of rice. *Bangladesh J. Botany. 18*: 67-72, 1989.

MULK, M.M. *Meloidogyne graminicola.* CIH Descriptions of Plant Parasitic Nematodes. Set 6, No. 87. *Farnham Royal, U.K. Commonwealth Agricultural Bureaux,* 1976.

ORTON WILLIAMS, K.J. *Meloidogyne javanica.* CIH Descriptions of Plant-Parasitic Nematodes. Set 1, No. 3. *Farnham Royal, U.K. Commonwealth Agricultural Bureaux,* 1972.

ORTON WILLIAMS, K.J. *Meloidogyne incognita.* CIH Descriptions of Plant-Parasitic Nematodes. Set 2, No. 18. *Farnham Royal, U.K. Commonwealth Agricultural Bureaux,* 1973.

OU, S.H. *Rice Disease.* Commonwealth Mycological Institute. Kew, England, 1972.

PADHI, N.N. & S.N. DAS. Biology of *Helicotylenchus abunaamai. Indian J. Nematol. 16*: 141-148, 1986.

PANDA, M. & Y.S. RAO. Evaluation of losses caused by incidence of *Hirschmanniella mucronata* Das in rice (abstract). *All India Nem. Symp.* New Delhi, India, 1, Aug. 21-22, 1969.

PRASAD, J.S.R. & Y.S. RAO. Ectoparasitic nematode fauna of upland rice soils. *Beitrage zur Tropischen Lond wirtschaft und Veterinar medizin, 22*: 79-82, 1984.

RAHMAN, M.L. & A.A.F. EVANS. Studies on host-parasite relationships of rice stem nematode, *Ditylenchus angustus* (Nematoda: Tylenchida) on rice, *Oryza sativa, L. Nematologica, 33*: 451-459, 1987.

RAHMAN, M.L. & S.A. MIAH. Effect of seedling establishment methods and carbofuran application to control ufra disease (*Ditylenchus angustus*) in deep-water rice. *Bangladesh J. Pl. Path. 5*: (in press), 1989.

RAHMAN, M.L. & S.A. MIAH. Occurrence and distribution of white tip disease in deep-water rice areas in Bangladesh. *Revue Nematol. 12*: 351-355, 1989.

RAMANA, K.V., S.C. MATHUR & Y.S. RAO. Role of *Hoplolaimus indicus* on the severity of seedling blight of rice. *Curr. Sci. 43:* 687-688, 1974.

RAMANA, K.V., & Y.S. RAO. Evaluation of yield losses due to lance nematode (*Hoplolaimus indicus* Sher) in rice. *Andhra Agric. J. 3:* 124-128, 1978.

RAO, Y.S. Physiology of resistance in rice to the root knot nematode. *Proc. 3rd Int. Cong. Pl. Path. Munich,* August 16-23, 153, 1978.

RAO, Y.S. Nematodes. *Rice in India,* ICAR. Monongr. Revised Ed. Padmanabham, S.Y. 591-615, 1985.

RAO, Y.S. Resistance in rice to parasitic nematodes. *1st Nat. Symp. on Genetics and Rice Improvement.* Hyderabad, Aug. 4, 1985.

RAO, Y.S. Current research on cyst and root knot nematodes of rice in India. *Advances in Plant Nematology, Proc. US-Pak. Internatn. Workshop on Plant Nema. (1986)*; 205-227, 1988.

RAO, Y.S. & H. BISWAS. Evaluation of yield losses in rice due to the root knot nematode. *Indian J. Nematol. 3*: 74, 1973.

RAO, Y.S. & A. JAYAPRAKASH. Leaf chlorosis in rice due to root infestation by a new cyst nematode. *Int. Rice Res. Newsl. 2*: 5, 1977.

RAO, Y.S. & M. PANDA. Study of plant parasitic nematodes affecting rice production in the vicinity of Cuttack (Orissa), India. *Final Tech. Rep. (USPL 480 Project).* Central Rice, Mimphis, Tennessee, 1970.

RAO, Y.S. & J.S. PRASAD. Root-lesion nematode damage in upland rice. *Int. Rice Res. Newsl. 2*: 6-7, 1977.

RAO, Y.S., J.S. PRASAD & M.S. PANWAR. Nematode problems in rice: crop losses, symptomology and management. *Plant Parasitic Nematodes of India, Problems & Progress* (Eds. G. Swarup & D.R. Dasgupta): 279-299, 1986.

RATHAIAH, Y. Diseases of deep-water rice in Jorhat District, Assam, India. *Int. Rice Res. Newsl., 14*: 32-33, 1988.

RAY, S., N. DAS & H.D. CATLING. Plant parasitic nematodes associated with deep-water rice in Orissa, India. *Int. Rice Res. Newsl. 12*: 20-21, 1987.

SAMSOEN, L. & E. GERAERT. Nematode fauna of rice paddies in Cameroon. I. Tylenchida. *Revue zool. Afric. 89*: 536-554, 1975.

SANCHO, C.L. & L. SALAZAR. Nematode parasitic on rice in southeastern Costa Rica. *Agron. Costa. 9*: 161-163, 1985.

SESHADRI, A.R. & D.R. DASGUPTA. *Ditylenchus angustus*. C.I.H. Descriptions of plant parasitic nematodes. Set 5, No. 64, *Farnham Royal, U.K. Commonwealth Agricultural Bureaux*, 1975.

SIDDIQI, M.R. *Hirschmanniella oryzae*. C.I.H. Descriptions of plant parasitic nematodes. Set 2, No. 26, *Farnham Royal, U.K. Commonwealth Agricultural Bureaux*, 1973.

SIDDIQI, M.R. *Tylenchorhynchus annulatus (T. martini).* C.I.H. Descriptions of plant parasitic nematodes. Set 6, No. 85, *Farnham Royal, U.K. Commonwealth Agricultural Bureaux*, 1976.

SIDDIQI, M.R. The origin and phylogeny of the nematode Orders Tylenchida Thorne, 1949 and Aphelenchida n. ord. *Helminth Abstr. Ser, B., 49*: 143-170, 1980.

SHER, S.A. Revision of the genus *Hirschmanniella* Luc & Goodey, 1963 (Nematoda: Tylenchoidea). *Nematologica. 14*: 243-275, 1968.

SINGH, B. Some important diseases of Paddy. *Agril. Anim. Husbandry, Uttar Pradesh, 3*: 27-30, 1953.

SIVAKUMAR, C.V. & E. KHAN. Description of *Hirschmanniella kaveri* sp.n. (Radopholidae: Nematoda) with a key for identification of *Hirschemanniella* spp. *Indian J. Nematol. 12*: 86-90, 1982.

STERLING, G.R. & J.S. McCULLOCH. *Paralongidorus australis* n.sp. (Nematoda: Longidoridae) causing poor growth of rice in Australia. *Nematologica, 30*: 387-394, 1984.

TAYLOR, A.L. Nematode problems of rice. *"Tropical Nematology"* (Eds. G.C. Smart, Jr. & V.G. Perry) Gainesville, University of Florida Press. 68-80, 1968.

TIMM, R.W. The occurrence of *Aphelenchoides besseyi* Christie, 1942 in deep-water paddy of East Pakistan. *Pak. J. Sci. 7*: 47-49, 1955.

TIMM, R.W. Nematodes parasites of rice in East Pakistan (abstract). *Proc. Pak. Sci. Conf. 12* (1960), Sec. B: 25-26, 1956.

TIMM, T.W. & M. AMEEN. Nematodes associated with commercial crops in East Pakistan. *Agric. Pakistan, 3*: 1-9, 1960.

VENKITESAN, T.S. & J.S. CHARLES. The rice root nematode in lowland paddies in Kerala, India. *IRRI Newsl. 4*: 21, 1979.

VUONG, H.H. The occurrence in Madagascar of the rice nematodes, *Aphelenchoides besseyi* and *Ditylenchus angustus. Tech. Comm. Commonw. Bur. Helminth. 40*: 274-288, 1969.

WAELE, D. DE. *Trichodorus petrusalberti* n.sp. (Nematoda: Trichodoridae) from rice with additional notes on the morphology of *T. sanniae* and *T. rinae. J. Nematol. 20*: 85-90, 1988.

WAELE, D. DE & E. VAN DEN BERG. Nematodes associated with upland rice in South Africa, with a description of *Hemicycliophora oryzae* sp.n. (Nematoda: Criconematoidea). *Revue. Nematol. 11*: 45-51, 1988.